Liberalism at Work:
The Rise and Fall of OSHA

Labor and Social Change
A series edited by Paula Rayman and
Carmen Sirianni

Liberalism at Work

The Rise and Fall of

OSHA

Charles Noble

 Temple University Press
Philadelphia

Temple University Press, Philadelphia 19122
Copyright © 1986 by Temple University. All rights reserved
Published 1986
Printed in the United States of America

The paper used in this publication meets the minimum
requirements of American National Standard for Information
Sciences—Permanence of Paper for Printed Library Materials,
ANSI Z39.48-1984.

Library of Congress Cataloging-in-Publication Data

Noble, Charles, 1948–
Liberalism at work.

(Labor and social change)
Includes index.
1. Industrial hygiene—United States. 2. Industrial
safety—United States. 3. United States. Occupational
Safety and Health Administration. 4. Capitalism.
5. Liberalism. 6. United States—Social policy.
7. Industry and state—United States. I. Title.
II. Series.
HD7654.N63   1986   353.0083'0289   86-1766
ISBN 0-87722-421-8 (alk. paper)

To Judith

Contents

[viii]   Contents

.

Tables and Figures

Table

Figure

Acknowledgments

A number of people helped me write this book. Some read drafts and commented on them; others lent the emotional support a project of this sort requires. I would like to thank in particular Patricia Bak, Stephen Bronner, Thomas Ferguson, Sandy Flitterman, Sam Friedman, Jim Hawley, Robert Kaufman, Carol MacLennan, Douglas Nelson, Lawrence Noble, Michael Rogin, Wendy Sarvasy, Roy Waldman, and Brian Wilson. Joel Rogers generously contributed his insight and knowledge in many hours of conversation. I owe him a special debt. I would also like to thank the members of the Rutgers University Political Economy Colloquium for the opportunity to try out some of the ideas for this book in their formative stages.

The research librarians and archivists at the National Archives, the Lyndon Baines Johnson Library in Austin, Texas, and the Jacob Javits Collection at the State University of New York, Stony Brook, gave me invaluable assistance in locating documentary materials. Anne Baylouny, Ruth Kenrick, and Deborah Blendowski helped with the research in New Brunswick. The reviewers, editors, and staff at Temple University Press were a pleasure to work with. In particular, Michael Ames's advice to make the argument as strong as I could made this a better book.

I would also like to thank the labor activists, unionists, public officials, agency staffers, corporate executives, and industry lobbyists who lent their time and insight to my effort to reconstruct this story. Most of them had other things to do, but nearly everyone I asked took time out to help. Some asked for anonymity, and I have respected their wishes. Those who did not are listed in an appendix to the text.

Finally, I want to thank Judith Grant, my colleague and companion. Her encouragement, good sense, and laughter helped me through the hard times and made the good times better.

Introduction

n 1970 Congress passed the Occupational Safety and Health Act (OSH Act) and committed the state to protecting workers from industrial accidents and occupational diseases. The Department of Labor was given the authority to set standards governing the working conditions of most American workers. It was also given the right to inspect workplaces and fine employers who violated those standards. In 1971 the department created the Occupational Safety and Health Administration (OSHA), an executive agency, to implement the act.

Today, despite the efforts of labor unions and health and safety activists to force public officials to fulfill this promise, little of lasting significance has been accomplished. OSHA has set few new major standards. The agency's inspectorate is small. Its programs have had little measurable impact on working conditions.

Why has OSHA failed to improve significantly the health and safety of American workers? Public opinion supports the effort. Americans strongly endorse the general aims of protective, or "social," regulation. Occupational safety and health regulation is particularly popular.[1]

The law itself gives workers impressive rights. The right to health and safety is expansive and broadly defined. The act's stated purpose is "to assure so far as possible *every* working man and woman in the Nation safe and healthful working conditions. . . ."[2] To accomplish this, the act requires that each employer "furnish to each of his employees employment and a place of employment which are free from recognized hazards that are causing or are likely to cause death or serious physical harm to his employees."[3] To enforce these rights, the secretary of labor is required, when issuing standards that deal with health hazards, to set standards that assure that "no employee will suffer material impairment of health or functional capacity even if such employee has regular exposure to the hazard dealt with by such standard for the period of his working life."[4]

In fact, the law is a remarkable piece of social legislation—radical in scope and vision in two important ways. First, the worker's right to protection is substantive: the state has a positive obligation to reduce risks. Second, this right is nearly universal; all but a few categories of workers (principally public-sector employees) enjoy it. In contrast, conventional labor legislation gives workers procedural rights. The Labor Management Relations Act (including the Wagner Act of 1935, the Taft-Hartley Act of 1947, and the 1959 Landrum-Griffin amendments) guarantees employees the right to form and join unions. The actual, substantive conditions of work are left to bargaining among employers and employees. Under these arrangements workers have to fight on their own for safe working conditions. The better organized may succeed; the unorganized or poorly organized live with what their employers choose to do. The state is officially indifferent to these outcomes. Its obligations are satisfied if employers do not prevent workers from exercising their rights to unionize.[5]

Regardless of this right to safe work, the OSH Act has not been implemented. Despite a positive obligation to protect workers from occupational accidents and diseases, administration after administration has balked at taking these rights seriously. As a result, workers remain at risk—exposed to over 2000 suspected carcinogens and the multiform accidents that can cut, crush, and maim the human body.

This book is about the state's failure to implement the OSH Act. It describes and explains how and why OSHA has been unable to provide workers with the kind of protection that Congress intended.

Based on this explanation, it proposes a way of restructuring the agency's approach to occupational hazards to improve its performance. Other studies have tackled aspects of this problem. In fact, given OSHA's short history, there is a rather large literature on it. But this book differs from other works in several respects. Most important, it proposes a radically different explanation for the agency's failures from those commonly found in both popular and scholarly accounts.

The failure to implement the OSH Act has been explained in one of three basic ways. Some suggest that the goals of the act are to blame; the law's right to health and safety is not economically feasible in a highly competitive international economy where minimizing the costs of production is so important to the success of firms. In short, it is impossible to provide workers with the blanket protection that Congress stipulated.[6] Others suggest that the approach taken by the OSH Act is flawed. Its emphasis on what has been called "command-and-control" regulation to change employers' behavior—that is, on detailed standards enforced by citations and fines—is misguided and counterproductive. Distant bureaucrats force firms to make changes that have little to do with the realities of production or hazard control.[7] Still others argue that OSHA has been frustrated by political opposition—by a business-backed movement for deregulation and, in recent years, the Reagan administration's intense opposition to government regulation of industry.[8]

Some of these points make sense; others do not. Clearly, political opposition has taken its toll on the agency. And the emphasis on detailed standards and penalty-based inspections has not proved particularly fruitful. The economic critique of the OSH Act's right to health and safety is less compelling. It fails to ask or answer why worker protection has borne so much of the brunt of economic decline, while other programs and policies, from defense spending to social security, have proven less vulnerable.

Even those points that do make sense provide partial or incomplete accounts of the agency's failure. Command-and-control regulation may be inefficient, but it is by no means obvious that it should prove so hard to implement. To the contrary, reformers rely on it because it is a relatively simple way of translating broad policy goals into practice. As for accounts that stress political opposition to OSHA, these fail to explain why that opposition has been so effective: why have public officials proven so responsive to the deregulation movement despite the support of public opinion, organized labor, and the environmental movement for occupational safety and health regulation?

A complete account requires a different perspective, one that locates the agency's failures in a broader, political-economic context. This is the perspective taken here. And from this perspective, other factors stand out. Most important, as I argue, there is a fundamental mismatch between the goals of the act and the approach taken to the control of occupational hazards. This mismatch extends beyond the choice of a command-and-control approach to occupational safety and health regulation to the basic ways in which workers, unions, social reformers, and public officials in the United States attempt to regulate markets and the process of capitalist production.

This choice, then, is part of a wider "liberal" mode of government intervention into the economy. By "liberal," I mean an approach to social reform in which state action is strictly limited by the property rights of private firms, and workers do not participate directly in efforts to reconstruct capitalist social relations. Public officials do not exercise significant control over production or investment, and workers do not take an active part in decision making in industry or in the implementation of public policies toward business. While this approach makes political sense in the American context, where a Lockean liberal ideology enshrines private property rights and workers are demobilized as a class, it is an ineffective response to the various ways in which American capitalism discourages social reform in general and workplace safety and health in particular. Thus, the problem of implementation is deeply rooted, indeed systemic in the sense that it arises out of a fundamental disjuncture between the ways in which Americans seek to solve social problems and the ways in which the wider socioeconomic system discourages social change.[9]

The Capitalist State and Social Reform

Given my intention to consider the problem of occupational safety and health from this perspective, I have chosen to work with an analytic framework that focuses on the relationship between American capitalism as a whole and social policy and facilitates the analysis of alternative approaches to reform. For the most part, this framework builds on the recent literature on the "capitalist state." For those who are not familiar with the basic concepts, I provide a brief summary of them and how they apply to the problem at hand.[10]

At the most general level, this literature is concerned with revising and rethinking Marx's view of democratic capitalism. This view is familiar and can be summarized briefly. Taken as a whole, Marx's writings provide an account of how the normal operation of a capitalist system of production leads to the political and economic subordination of workers as a class. Marx argued that this subordination is structural, that is, it grows out of the way that production is organized. That structure is characterized by the concentration of productive assets in the hands of capitalists who, based on their property, appropriate and invest a socially produced surplus in the form of profits. This makes them dominant at the workplace, where employers dictate the conditions of work to employees; in the market, where workers are forced to sell their labor power at less than the value of what it produces; and in the state, where accumulated wealth and social position give capitalists political power over workers. Moreover, Marx argued, as long as production remains organized in this fashion, the subordination is inevitable. If workers want to be free, they have to change the system.

Generally, I find this view compelling. I think that capitalist democracies are class systems and that political and economic power are concentrated in the hands of private firms and investors. Moreover, democracy, at least as Marx defined it—a society in which freely associating producers govern themselves in all aspects of social life —requires a fundamental change in the underlying mode of production. The power of employers at work, and that of business in politics, discourages self-determination in too many ways to allow for democracy in the strong sense.

Nonetheless, as Marx's critics and many of his followers have noted, this view leaves students of reform with one very critical issue unresolved. In capitalist democracies workers enjoy legal rights to vote for and elect political representatives. Why cannot they use these rights to make the state *autonomous*, that is, capable of acting in the interests of workers and against the interests of business?

This was not a particularly important problem for Marx. Writing before the advent of full universal suffrage and the institutionalization of social democratic and labor parties, he focused on how conflicts between the bourgeoisie, aristocracy, and state bureaucracy shaped policy. There was little question that the "bourgeois" state served ruling-class interests. The issue for Marx was which interests. In Marx's historical accounts, social reform is largely driven by conflicts between competing ruling classes. Since the bourgeoisie was the ascendant class in the mid-nineteenth century, reform ultimately served to maintain its interests: the rule of capital.

That Marx held this view is clear in his discussion of English factory legislation. Ostensibly a victory for social reformers and the working class, factory reforms were, according to Marx's account, dictated by the logic of capitalist accumulation. Employers tended to so exploit workers that they imperiled the survival of the labor force that was the source of their profits. Therefore, Marx wrote, a state "ruled by capitalist and landlord" resorted to "forcibly limiting the working day by state regulations" in order to "curb the passion of capital for limitless draining of labor-power." Because capitalists needed workers if they were to continue to appropriate surplus value, the state protected workers, even against the industrialists' opposition to reform. Marx acknowledged the role that worker protest played in precipitating reform, but the protest was a background condition rather than a driving force.[11]

Lenin was forced to take social reform more seriously. The Bolsheviks were in competition with social democratic parties for the allegiance of workers, and he developed a more complete account of nonrevolutionary political strategies. For Lenin, the liberal democratic state was destined to serve capitalist interests by a multiplicity of factors. Its form, for example, precluded direct participation in decision making by workers; both parliament and the bureaucracy were distant organs hostile to popular control. The executive branch's financial dependence on capitalists to fund the public debt further limited its autonomy. State bureaucrats themselves were either tied to capitalist interests financially or ideologically predisposed to favor them. Economic interests also dominated the political parties that organized the legislature. Finally, because the rule of law created a false impression of universal rights, it obscured the class nature of the state and made it difficult for workers to understand or act on their class interests. In fact, to Lenin, the democratic state was "the best possible political shell for capitalism."[12]

Much of Lenin's argument is to the point, particularly his emphasis on the role of direct participation, or the lack thereof, in determining the ends of state action, and the importance of the state's economic dependence on private investors to fund the public debt. But Lenin's commitment to discrediting his social democratic opponents led him to overstate his case. Most important, he treated all these factors as invariant, despite the fact that all can be influenced by working-class political action.

Recent work on the capitalist state and social reform has been much more successful with dealing with the contingent character of class rule. Comparative and historical research has demonstrated that capitalist states do act against the professed and "objective" interests of capitalists and that this occurs in large part because workers organize and are able to affect the balance of power between themselves and business. State theorists and political economists have taken these findings to heart and have developed more complex ideas about the relationships among the structures of production, political activity and organization, and public policy in capitalist democracies.

Two ideas in particular have helped illuminate the contingent character of business power. The first suggests that we view the capitalist state as "constrained," rather than "determined," by the structural organization of the political economy. The second suggests that we consider the conditions under which workers choose to consent to capitalism rather than challenge its basic socioeconomic institutions.

The idea of constraint suggests that the state is neither directly governed by business nor required by its location within the larger system or its form to serve only business interests. Instead, public officials are encouraged to act in this fashion by the incentives they face. If these incentives are changed, public officials can lead the state in different directions.

The constraints on the state are predominantly structural and are based in the double dependence of the state on the process of capitalist investment. First, the state's ability to act in domestic or international affairs depends on its ability to raise revenues through taxation. This depends on a healthy economy, which in turn depends on the level of private investment. Second, because workers vote their pocketbooks, the political fortunes of elected leaders depend on the level of economic activity. When jobs and incomes are insecure, parties fail at elections and administrations fall. Thus, in order to maintain the state's power and secure their own political fortunes, public officials seek to maintain "business confidence" in their administrations. Even if they are sympathetic to labor, they avoid policies that threaten capitalist profitability and choose policies that serve business interests.[13]

The focus on the conditions under which workers choose to consent to capitalism highlights a second face of business power and further illuminates its contingent character. According to this view, the economic structure of capitalist democracy encourages workers, like public officials, to act in ways that are compatible with the interests of business.

Most important, capitalist democracy *discourages* radical collective action by workers. It does this in three ways: first, the concentration of productive assets in private firms and investors leaves workers dependent on employers for jobs and income. Workers recognize that their economic interests depend on the prior satisfaction of business interests and "rationally" limit their demands to those things that are compatible with firm profitability. Second, the capitalist organization of the labor process creates and reinforces preexisting divisions among workers, including divisions based on skill, race, gender, and ethnicity. This divisiveness makes coordinated, solidaristic collective action difficult. Third, resource inequalities between workers and those who own and manage capitalist enterprises make it difficult for workers to pursue political activities of all sorts. Unequal access to free time and differences in financial resources and information make it hard for workers to establish the organizational preconditions for effective radical action. Repeated failures, or even the likelihood of failure, further encourage workers to confine their demands to more easily achieved ends.[14]

In these ways, the structure of capitalist democracy encourages workers to consent to class domination. Workers, calculating the costs and benefits of various forms of political and economic activity, choose political strategies that emphasize short-term material gain, including job security, higher wages, shorter hours, and changes in the most immediate conditions of work. The economic insecurity of the wage earner, divisions among employees, and employer strategies to contain potential opposition make the costs and risks of more radical collective action too high.

Taken together, these efforts to clarify the contingent character of class rule have identified three critical variables that affect the ability of capitalist states to act in the interests of workers and against the interests of business: (1) the degree to which workers are organized and capable of acting as a class; (2) the degree to which the state is capable of independently affecting the flow of capital to alternative uses; and (3) the degree to which public officials adopt regulatory programs that actively involve workers in the implementation of reforms.[15]

The idea of rational consent suggests that the political and economic strategies of workers also shape the state's freedom to act for or against business. The existence of an independent party political force representing workers as a class, extensive unionization, centralization in the trade union movement, and the pursuit of radical political rather than short-term economic goals are likely to help free public officials from their dependence on capitalist investment. It follows that if the state is constrained by its structural location in the wider political economy, it can increase its ability to act against the expressed interests of business if it is able to loosen its economic dependence on capitalist investment. Finally, workers can be encouraged to demand more radical political changes if they are included in the implementation of public policy and decision making at the workplace.

These observations provide the framework for the analysis of social reform in this study. They suggest that whereas the limits of reform originate in capitalist economic structures, they are sustained only to the degree that political strategies fail to confront those structures, most important the private control of investment and the exclusion of workers from participation in the organization of work and the implementation of policy. To the extent that these structures are left in place by reformers, social policies that depend for their success on radical political action by workers, or on anticapitalist political action by public officials, are likely to fail.

Alternatively, reforms that increase public control over investment are likely simultaneously to increase the willingness of public officials to challenge business interests and the willingness of workers to choose more radical political strategies. Reforms that widen the scope of worker participation in the organization of work are also likely to encourage workers to act more radically by facilitating collective action. In sum, reforms of this kind widen the field for independent state action.

Whether independent state action follows, or finally succeeds, of course depends on a host of other factors. Public officials may have the opportunity to challenge the existing order but be committed to it ideologically. The costs of radical political action for workers may decrease substantially, but labor unions and worker parties can still choose not to respond affirmatively. Procapitalist ideologies may obscure, or devalue, the possibility of greater worker power and participation within the system. Or party leaders may fail to understand and respond to these opportunities. Nonetheless, the strategy of reform matters significantly in determining what social policy can or cannot accomplish.

Given this analysis, American liberalism appears particularly infirm as an approach to reform. As a rule, when reformers succeed, they tend to create public programs that rely on methods that concentrate and centralize authority in distant professionalized bureaucracies rather than extend public authority over investment or involve affected constituents actively in the implementation of policy. Without greater public control over investment, the ability of elected officials to take actions that threaten business confidence is sharply restricted. In the absence of institutions to facilitate worker participation in the governance of firms and citizen participation in a host of other basic economic decisions from plant location to community health services, regulatory agencies find it difficult to significantly alter the distribution of power and property in American capitalism.

The problem, however, is not simply one of bureaucratic organization. To the contrary, the weakness of the labor movement and its preferred political and economic strategies helps to reproduce liberalism in policy. Most workers are not organized and those that are do not belong to classwide organizations, either in industry or in politics. There is no central labor federation that speaks for the majority of American employees, no labor party to represent worker interests in politics. Nor, for the most part, does organized labor demand greater public control over capital accumulation. Rather, American unions have ceded control over production and investment to private firms and managers and, within this frame, sought economic security and higher standards of living. In fact, the American labor movement is single-mindedly "economistic" in its pursuit of short-term material gain and distinctive in the degree to which it eschews more radical political and economic changes.

The Political Economy of Workplace Regulation

The framework I have outlined can be applied directly to the problem of working conditions. It suggests that occupational safety and health policy will succeed only to the extent that it enables public officials to challenge those aspects of the economic system that make jobs unsafe and discourage private and public actions to improve working conditions.

Chapter 1 explores these issues in depth. There I suggest that in industrial capitalist societies, most accidents and diseases result from the decisions of private employers and private investors. Workers may contribute by acting in an unsafe manner, but they work in environments structured by the decisions of firms to invest in certain products and technologies and by employers' decisions to organize work in certain ways. Hazard control rests on changing these decisions so that firms increase their investment in health and safety and workers participate actively in plant-level decision making over the conditions of work.

Capitalism tends to discourage both. Profit-seeking investment in competitive markets tends to discourage investment in health and safety. The capitalist organization of the labor process discourages worker participation in determining the conditions of work. Although both are essential to effective hazard control, both run counter to employer interests. Finally, the state's dependence on private production discourages public officials from interfering with managerial prerogatives for fear of alienating employers and undermining business confidence. Chapter 2 illustrates how, in near-pristine form, these processes discouraged prevention in the United States before 1970.

These processes do not preclude the creation of regulatory mechanisms to reform the conditions of work. Under the right conditions, government may prove receptive to demands for workplace regulation. As Chapter 3 indicates, the combination of economic growth and social protest in the 1960s encouraged public officials to respond affirmatively to demands for federal intervention. Indeed, the passage of the OSH Act in 1970 over the concerted and vehement opposition of the business community confirms that the capitalist state can be receptive to worker demands.

Still, the theoretical observations outlined above suggest that successful reform remains problematic unless reformers and the labor movement directly address the constraints that circumscribe the independence of public officials and discourage workers from taking a more active role in determining the conditions of work. This did not happen here. Neither the act nor union strategies toward work took this approach.

Instead, the law and efforts to implement it have remained resolutely liberal and, thus, have failed to confront the impact of market capitalism on occupational safety and health. Rather than extend public authority over the economic processes that structure the workplace, or redefine the role of workers in determining the conditions of work, the law and efforts to implement it have relied on command-and-control regulation to limit worker exposure to hazardous work. Chapter 1 places this approach in perspective by comparing liberal forms of regulation in America to neocorporatist forms of regulation common in Europe. Based on that comparison, it offers a preliminary theoretical assessment of the strengths and weaknesses of liberalism as an approach to the control of occupational hazards.

In some respects, the OSH Act did attempt to reshape the American approach to workplace regulation. Several provisions provide for worker participation in enforcement. One mechanism was created to involve employer and employee organizations in negotiations over standards. Taken as a whole, however, the act deepened rather than altered the state's commitment to a liberal mode of reform. Rather than reorient the existing system, the state-level programs were federalized and command-and-control enforcement mechanisms were strengthened. Chapter 3 explains why Congress and the executive branch opted for this and not an alternative approach.

The general account of the capitalist state and social reform suggests that this approach to occupational safety and health should prove difficult to implement. The state's inability to regulate investment leaves public officials vulnerable to declining business confidence and significantly increases employers' influence over social policy. The failure to involve workers in enforcement deprives OSHA of the additional leverage that a mobilized workforce could exert on employers and in politics. As a result, the agency is likely to be subjected to enormous political pressure from business and from elected officials whom business groups hold responsible for the agency's policies. Agency rulemaking and enforcement will become highly politicized. In this environment, the agency will make policy in response to short-term political forces rather than design and implement long-term hazard-control programs. Rational policymaking will suffer.

All of this has occurred, as Chapters 4 through 7 describe. After 1970, employers targeted by OSHA actions sought relief from the agency, Congress, and the White House. As economic conditions and profit margins declined in the 1970s, employers intensified their opposition to the agency. At the same time, economic instability made public officials sensitive to their demands. Worker rights to protection quickly became vulnerable to counterclaims based on the importance of economic growth and capital investment.

Presidents proved particularly sensitive to business demands. This was the result, in part, of White House sensitivity to organized pressure. But it also reflected presidential interest in sustaining business confidence in a time of economic crisis. The executive branch responded to both concerns by extending central control over the agency. The precise form and content of White House review reflected the changing balance of political forces in each administration. Nevertheless, a common trajectory is apparent in the efforts of Presidents Gerald Ford, Jimmy Carter, and Ronald Reagan to increase executive oversight over OSHA and introduce economic criteria into the agency's decision-making processes. As I indicate, these programs played an important part in shaping what OSHA could and could not do.

Many factors account for the success of business opposition to OSHA in the 1970s and 1980s. The ability of business to outspend organized labor in elections and to mount better-funded lobbying efforts on Capitol Hill clearly contributed. No doubt the conservative drift in the political climate made it easier for industry to be heard. But changes in political strategy were also critical. Chastened by a string of legislative defeats in the 1960s and early 1970s—particularly on air and water pollution, occupational safety and health, and consumer product safety—business groups mobilized in new ways. They built organizations that could represent their common class interests, and they rehabilitated their ideology to offer a new defense of their interests, based on the link between firms' profits and society's interest in economic growth. Eventually, the business lobby took back the initiative it had lost in the 1960s and forced the health and safety movement onto the defensive.

Union strategies also contributed to OSHA's troubles. The labor movement lobbied for enforcement of the OSH Act but did not seek to mobilize workers in politics or at the workplace. Nor did it propose a new vision of work, or a democratic economy, that could link the interests of workers in workplace reform to the demands of the civil rights, environmental, consumer, and feminist movements for a safer, more egalitarian, more participatory society. Unions formed alliances with these groups, but, on the whole, organized labor continued to lobby in conventional forms, pressuring the agency for standards that protected particular groups of workers. As a result, the agency was forced to deal with a profoundly asymmetrical political environment in which the balance of power shifted toward employers.

Chapter 7 looks at how OSHA responded to these political forces. Organizational theory suggests that OSHA should have sought autonomy from political pressure and crafted its policies and internal decision-making processes to maximize its discretion and resources. In its early years it should have attempted to establish itself as a forceful actor in its environment by adopting strong rules and defending them. Subsequently, it should have sought to minimize conflict by adopting a low profile, courting the best-organized interest groups in its strategic environment.[16]

But politics, not organizational imperatives, was in command. Competing demands by organized labor and business, in combination with the White House review programs, so politicized the agency's environment that it was never able to establish itself as an independent voice. Instead, its policies constantly changed as the balance of political power shifted from administration to administration and, in some cases, from year to year. Chapter 7 considers how these changes shaped OSHA's approach to job safety and their impact on working conditions. As the theoretical analysis suggests, the agency was able to do little to improve worker safety and health. Instead, competing political demands led to inconsistent and often irrational policies. After 1980, business and White House pressures led to near-total deregulation.

Regulatory Reform

OSHA's failures are well known, and there are numerous proposals to restructure the agency. The two most common approaches to regulatory reform are (1) a market-conservative proposal that argues for deregulation and greater reliance on employer discretion and the operation of labor markets; and (2) a neoliberal proposal that recommends centralized economic review of health and safety regulation to make standards and enforcement more compatible with the needs of economic growth and international competition.

Neither of these conventional proposals takes into account the structural and institutional factors considered here. Chapter 8 critically evaluates them according to the theory and evidence presented in this book. It also considers and argues for a third route to regulatory reform that is designed simultaneously to increase worker participation at work and increase public authority over employer decisions that directly affect investment in health and safety.

This third way does not simply recommend a stronger OSHA; it approaches the entire problem of regulating workplace hazards differently. It suggests that the state adopt policies that make it possible for workers to become less dependent on OSHA and more reliant on their own political and economic organizations to change the conditions of work. The state continues to play an important role, but the focus of public policy shifts from detailed standards and penalty-based inspections to creating state-enforced rights to participate in plant governance. This way also requires that labor strategies change. Unions must play a more active role at the national level in setting basic occupational safety and health policy. They must work to increase public control over investment in order to loosen the structural constraints on public officials. They must work to focus worker activity at the local level on the conditions of work. And they must build alliances with other movements interested in participatory approaches to economic decision making and policy implementation.

The Limits of Liberalism

Beyond the theory of social reform in capitalist states, and the analysis of occupational safety and health regulation, I intend this book to provide the foundations for a critical account of liberalism as a strategy of reform in America. As in the case of OSHA, other Great Society programs combined new, quite radical rights with highly statist enforcement programs, including air and water pollution control, consumer protection, and civil rights. Thus we can consider what OSHA suggests about these more general trends.

Of course, there are limits to what can be learned from a single case. Theories cannot be validated or invalidated by one example. It is difficult to analyze an approach to reform on the basis of a single issue. And OSHA is distinctive in several respects: the act's rights to health and safety are particularly expansive; most welfare state policies rest more heavily on government's taxing and spending powers than on command-and-control regulation. But there are sound methodological reasons to endorse the case method. And the parallels between OSHA and the general trajectory of the Great Society are sufficiently strong to justify the effort.

OSHA's critics agree; they also see the agency as emblematic of deeper dynamics in American politics. To conservatives, OSHA is probably the nation's leading symbol of overregulation. Its history is a cautionary tale about what is wrong with liberal state intervention. Conservatives suggest that OSHA's failure reflects a general tendency for bureaucratic agencies to fail to accomplish their legislatively set goals. Bureaucracies adopt policies that serve their organizational interests rather than solve social problems. As a result, rational policy-making suffers. In contrast, left liberals and radicals suggest that the history of workplace regulation illustrates how the political power of business in American politics frustrates reform.

The conclusion returns to these issues and suggests that the problems of liberalism in the case of OSHA are characteristic of the basic avenue taken to social reform in America. Specifically, I argue that liberalism is at once overly statist and insufficiently radical to solve the problems it confronts. As this study indicates, liberalism does not prevent people from demanding, presidents from endorsing, and Congress from adopting laws that promise to advantage workers and consumers and impose substantial costs and controls on firms. Liberalism does, however, build from a set of political and institutional assumptions that frustrate reform.

Two problems stand out. First, the state's effort to correct market "failures" such as pollution or workplace hazards without using public authority to affect the levels and kinds of capital investment, the industrial structure, or the mix of jobs offered leaves these policies highly vulnerable to political opposition and market forces. In an unstable economic climate, social reform is likely to be held hostage to the demands of firms and investors, demands that preclude the restructuring of markets to serve worker or consumer interests.

Second, liberal forms of state intervention are rarely joined to participatory mechanisms that enable the people who are to be protected to protect themselves. Instead, by relying so heavily on direct state supervision of firms—in this instance on command-and-control regulation—policy encourages the affected groups to transfer responsibility to public officials, the same public officials who are likely to be so vulnerable to business opposition.

This approach makes sense in the American context. It reflects a cultural preference for the least intrusive forms of state intervention and the political reality that workers and consumers are already demobilized at the workplace and in their communities. Nonetheless, it does not address the political and economic realities of the problems it proposes to solve. Although it creates a more extensive state, it does not substantially increase the state's effective power. In sum, liberalism fails to confront how capitalist democracy in general, and

American capitalism in particular, discourage reform. It follows that
any serious effort to remedy problems of this sort requires breaking
with this reform strategy. Participation and increased public authority
over economic processes are the preconditions for successful reform.
The right to protection depends on greater public control over production. At the same time, state action should be used to facilitate
self-organization by workers and consumers. Anything less radical is
likely to fail.

[1]
The Political Economy of Workplace Regulation

n capitalist societies, most occupational accidents and diseases arise in the process of private, profit-seeking production. Employers decide what and how to produce, and these decisions determine which technologies are adopted and how workplaces are organized. Workers, in turn, encounter or avoid hazardous work by moving in and out of risky jobs as they sell their labor in markets.

In industrial capitalist economies, the costs of protecting workers from health and safety hazards are high. In the United States, for example, the Council on Wage and Price Stability has estimated that moderately strict standards for all 2415 known or suspected carcinogens could eventually entail $526 billion in capital and recurring costs. OSHA has estimated that changes in machinery and plant design to reduce the permissible level of noise in American industry from 90 dBA to 85 dBA could lead to one-time capital costs of $18.5 billion and recurring annual costs of $1.1 billion. The combined costs of comprehensive chemical labeling to protect workers who handle toxic substances may be as high as $8.2 billion. No doubt some of these figures are inflated; exaggerated cost estimates have been used to discredit occupational safety and health regulation, as Chapters 4 and 6 indicate. Nonetheless, worker protection will absorb a substantial share of society's resources and significantly increase the costs of production.[1]

Left to its own devices, the economically rational firm should try to avoid incurring these costs. Although there are some benefits to be had from a safe workplace and a healthy workforce, least-cost methods of production argue for leaving workers and public officials to deal with the economic, psychological, and physical impact of accidents and injuries. The structure of capitalist democracy, in turn, discourages public officials and workers from attempting to alter this arrangement.

Given these political and economic processes, the ability of workers and public officials to improve working conditions depends on the adoption of a regulatory approach that increases their power vis-à-vis employers and facilitates challenges to least-cost principles of capitalist production and unregulated labor markets. This chapter provides a preliminary view of the range of available regulatory alternatives and a discussion of their comparative strengths and weaknesses.

By anchoring the analysis in the capitalist organization of work, I do not mean to suggest that occupational safety and health hazards are unique to capitalism. To the contrary, work can and has imperiled the lives of workers throughout history and across modes of production. Dusts from naturally occurring elements or minerals such as arsenic and asbestos have plagued people since they began working with them. Farmers and hunters, mechanics and laborers—all have had to deal with threats to muscle, tissue, and bone since they began to transform nature through work.

Industrial production has undoubtedly increased the number, kinds, and severity of occupational hazards. But capitalism is not alone in promoting industry or exposing workers to the hazardous by-products of industrial production. Socialist societies have not eliminated mine accidents or worker exposure to toxic chemicals. Soviet workers, for example, were exposed to polyvinyl chloride in the manufacture of plastic. Under Mao Zedong, Chinese workers were taught that strict observance of safety measures was "bourgeois" and "cowardly."[2]

Nor are all industries and firms equally unsafe. Data collected by the Bureau of Labor Statistics indicate that the all-industry private sector incidence of accidents and injuries per 100 full-time workers in 1980 was 8.7. On this measure, wholesale and retail trade (7.4), services (5.2), and finance and real estate (2.0) are relatively safe (i.e., below the all-industry rate). Conversely, construction (15.7), manufacturing (12.2), agriculture, forestry, and fishing (11.9), mining (11.2), and transportation and public utilities (11.1) are relatively hazardous (i.e., above the all-industry rate). The relative hazardousness of industries varies within these broad classifications, particularly in the manufacturing sector (see Table 1.1). Within industries, some employers are known to spend a great deal of time and effort on health and safety; others are more sanguine about occupational hazards.[3]

Nevertheless, nearly all American firms do share one critical thing in common: they are capitalist enterprises. As such, they create and distribute occupational hazards in similar ways. In market capitalist societies, economic decisions by private investors and corporate managers and, secondarily, wage workers determine the allocation of risk and protection. In contrast, in state socialist societies, political decisions determine the levels and kinds of hazards that workers are exposed to by establishing investment plans, production schedules, work rules, and allocating labor among jobs and occupations. In these societies, occupational safety and health hazards begin with centralized economic decisions. In capitalist societies, hazards emerge from uncoordinated, decentralized, profit-oriented decisions that structure the environment in which workers seek to protect themselves. If they hope to be effective, workers must adopt strategies, and the state must adopt forms of intervention, that are appropriate to this mode of production.

Capitalism and Occupational Safety and Health

The processes of investment and technical change are central to productive activity in all modern industrial societies. Industrial capitalism, however, accomplishes these twin tasks distinctively. The economic surplus available for investment appears in the form of profits, and profit-seeking investors initiate economic growth by commiting capital to production. Corporate managers, in turn, organize and reorganize the labor process to make production as profitable as

Table 1.1. Occupational Injury and Illness Rates of Manufacturing Industries, 1980

Industry	Incidence Rate
Food products	18.7
Meat products	28.0
Lumber and wood products	18.6
Logging camps and contractors	22.7
Wood building and mobile homes	25.7
Fabricated metal products	18.5
Fabricated structural metal products	22.5
Furniture and fixtures	16.0
Rubber and plastics	15.5
Primary metal industries	15.2
Iron and steel foundries	23.6
Stone, glass and clay products	15.0
Nonelectrical machinery	13.7
Paper products	12.7
Leather products	11.7
Leather tanning and finishing	23.5

possible. We can argue over whether profits are the only goal that corporate managers and investors pursue. But since profit is both the necessary and limiting condition for the pursuit of all other goals, I take profit maximization by private investors and corporate managers to be the starting point of the analysis in this section.[4]

Employer Strategies Toward Work

In the process of production, profit maximization leads employers to adopt two related strategies. First, employers attempt to maximize their control over the workplace in order to obtain the greatest pos-

Table 1.1.—Continued

Industry	Incidence Rate
Transportation equipment	10.6
Ship and boat building and repairing	25.6
Tobacco products	8.1
Electric and electronic equipment	8.0
Petroleum and coal products	7.2
Printing and publishing	6.9
Instruments and related products	6.8
Chemical products	6.8
Apparel and other textiles	6.4
All Manufacturing	12.2

Note: The incidence rates represent the number of injuries and illnesses or lost workdays per 100 full-time workers. General groupings refer to 2-digit Standard Industrial Classification (SIC) codes. Subgroups refer to 3-digit SIC codes. This list does not include all 2- and 3-digit industries.
Source: Bureau of Labor Statistics, *Occupational Injuries and Illnesses in the United States, 1980* (Washington, D.C.: GPO, 1982), table 1, pp. 2–13.

sible work effort from their employees. Second, they seek to minimize the costs of production.[5] Ordinarily, both strategies are likely to increase the hazards that workers are exposed to and discourage efforts to promote safe and healthy work.

In combination, these strategies can inhibit workplace safety and health in several ways. In attempting to maximize the amount of effort that workers devote to their jobs, employers are led to devalue work practices that make jobs safer. For example, speeding up the production process or reducing the number of workers assigned to each task without reducing production quotas are the simplest ways to intensify work. Both increase the likelihood that workers will make mistakes, be hurt by machinery, or be exposed to other hazards.

Cost minimization also undermines safe working conditions. The economically rational manager adopts new technologies that heighten labor productivity and lower raw materials costs. New product lines that increase market shares are likely to be favored. Many new materials, products, and production processes are hazardous, however. Often the hazards are not known, or if known, methods to control them are not fully developed. But without some mechanism to prevent employers from taking advantage of these investment opportunities, they are apt to develop new products and exploit new markets with less than full regard for the impact of these decisions on working conditions.

Cost minimization commonly entails forgoing safe practices that absorb time and money that otherwise could be saved or devoted to more economically productive uses. Safety tends to raise the costs of production. Two workers may be able to perform a task more efficiently than three; but three workers are likely to do it more safely. Presumably, this "redundant" labor, though adding to workplace health and safety, will be cut in the interests of profit maximization.

Engineering controls, that is, changes in the technical organization of the production process, are generally more effective means of controlling hazards than what specialists in the field call *personal protective devices* (PPDs), such as dust masks, hard-toe shoes, and ear plugs. Engineering controls are also considerably more expensive than PPDs. Understandably, employers prefer to avoid them whenever possible. Serious worker health and safety training programs are also expensive: they take time away from work; money must be spent on program development and instruction. These efforts, too, are likely to be sacrificed when profitability is at stake. Finally, except when they are convinced that hazardous conditions threaten to damage plant and machinery (as, for example, refinery explosions do), employers are reluctant to shut down production processes or remove workers from risky jobs.

Less generally recognized but equally important, employers' efforts to maximize control over the workplace and minimize costs tend to discourage worker participation at the workplace, and this further undermines occupational safety and health. Research indicates that worker participation in plant safety and health programs and policy-making has a salutary impact on the conditions of work. It provides managers with immediate input about potential hazards and imminent dangers; it encourages workers to cooperate in safety efforts; it incorporates workers' firsthand knowledge and experience in the definition and solution of health and safety problems.[6]

Nevertheless, taken seriously, worker participation conflicts with managerial control over the labor process. Effective participation means that workers exercise a certain degree of autonomy at work and cooperate among themselves. Both facilitate employee resistance to managerial directives. Therefore, employers who seek to maintain and augment their control are likely to develop highly bureaucratized and centralized forms of personnel management. They will divide production processes into highly differentiated job structures, rotate workers among different jobs, and limit the time and opportunity available to workers to form affinity groups that might provide the basis for worker opposition to managerial control. In collective bargaining, firms will insist on managerial prerogatives over plant administration issues. They will view challenges to the decisions and practices of plant administrators as challenges to managerial control of the labor process.

Finally, efforts to maximize work effort will make work more onerous and disagreeable, and the firm will be forced to supervise workers even more intensively. This will further exacerbate conflict and will encourage adversarial relationships between employers and employees. Worker participation would suffer, and with it, occupational safety and health; the adversarial atmosphere will vitiate attempts to involve workers in plant-level activities. Workers will be discouraged from voicing their concerns over working conditions or refusing hazardous work for fear of employer reprisal. They will also be more likely to view employer efforts to educate them about safety as company propaganda. Employers, in turn, will be predisposed to view worker complaints over working conditions as challenges to managerial prerogatives, and they will be reluctant to admit the existence of unsafe conditions.

In sum, the capitalist organization of production encourages firms to adopt strategies to intensify work and minimize the costs of production. These strategies, in turn, lead directly and indirectly to increased workplace hazards and discourage efforts that might otherwise be made to reduce the risks of work.

Why Employers Might Voluntarily Protect Workers

A number of countervailing tendencies make voluntary efforts on the part of employers to reduce occupational hazards economically rational. For example, accidents and injuries interrupt production, hurt skilled workers, lower productivity, and force employers to pay increased medical bills. These either raise the costs of production or lower total output. In any event, like wages and the costs of raw materials, they cut into employers' profits. In contrast, safe workplaces may increase employee productivity by increasing employee morale.

There also seems to be an inverse relationship between money wages and job hazards.[7] Some employers pay what are called *risk premiums*, that is, wage premiums designed to induce employees to take risky jobs. Conversely, some workers will accept a lower money wage if they are assured of relatively safe working conditions. The existence of these premiums should lead some firms to improve workplace safety in order to lower their wage costs. Individual firms also have an interest in safe work because it can facilitate managerial control of the workplace. Unsafe work breeds alienation and militancy and encourages wildcat strikes and other forms of employee protest.

Employers, as a class, also have common political and economic interests in job safety. Politically, employer programs to promote occupational safety and health can help create the impression that employers are socially responsible and augment their influence in politics. Economically, employers have an interest in conserving labor. A large number of disabled workers decreases the supply of labor; a plentiful labor supply keeps wage costs from rising. In fact, there is a corporate-sponsored movement that promotes workplace safety and health. Represented by the National Safety Council (NSC), it argues that "safety pays." Many of the safer firms and industries are involved in it, as we see in a later chapter.

But the disincentives to devoting resources to prevention are also great. Output can be increased by intensifying work. Market shares can be augmented by the quick introduction of new technologies whose long-term effects are unknown. Workers seldom have a good deal of accurate information about the hazards they face, and they are not likely to be able to rank competing jobs according to their relative risks. This is particularly true with health hazards, where the effects of exposure to toxic substances may take many years to show up, and where it is often difficult to establish the precise etiology of a disease. In addition, workers generally do not know what to make of the information they have. People have a hard time assessing the dangers of low-probability risks; they usually unduly discount these dangers.[8] This tendency is probably pronounced when workers are financially dependent on their jobs and have few marketable skills or economic options.

As for risk premiums, the evidence suggests that they do exist in some high-risk blue-collar occupations. But it is not clear whether there are risk premiums for nonfatal as opposed to fatal injuries, and it is far from established that these wage differentials provide significant incentive to employers to take greater care, particularly when hazard control is expensive.[9] As employers point out, hazard control can be very expensive.

Safer work may discourage labor militance, but other factors affect what workers do about occupational hazards. Where unions do not focus on health and safety, employers need not worry about strikes over working conditions. In any event, employers have other ways of imposing labor discipline and undermining the efficacy of shop-floor protests.

Finally, firms may have a collective, "objective" interest in conserving labor, but few mechanisms are capable of turning that interest into classwide policy. To the contrary, in the competitive marketplace, firms act as individual economic actors. As such, each firm can be expected to make its own calculations about the costs and benefits of voluntary hazard reduction. These estimates vary widely for different firms. Some companies are in relatively safe industries, such as communication or financial services. Other industries, (e.g., mining) include a large number of high-hazard jobs. Some firms work with old capital equipment that does not have the best available engineering controls. Others have safer equipment because they have retooled. Burlington Mills, for example, recently completed a capital program that substantially increased its ability to control cotton dust. Other smaller, less competitive textile firms have much more primitive equipment. Some companies have been forced to provide safer work by unions; the United Auto Workers (UAW), for example, has been relatively successful in forcing employers to reduce noise levels. Non-unionized manufacturing companies do not face pressure of this sort. We cannot rule out the possibility that class-conscious leadership can emerge to articulate a collective capitalist interest in occupational safety and health and attempt to impose it on business at large. Yet this rarely happens. When corporate leaders have argued for reforms of this sort, they have found it difficult to impose them on a diverse and competitive business community.

Forms of State Intervention

In response to occupational hazards, capitalist states are faced with a choice among three basic forms of intervention: (1) legal liability; (2) workers' compensation; and (3) factory legislation. These three forms are not mutually exclusive. They do, however, represent different governmental responses to the problem. They can be distinguished according to how they approach hazard control and how they respond to the capitalist organization of work.

Legal Liability

The legal-liability approach allows workers to use tort law to sue employers for damages from injuries sustained as a result of negligent exposure to harmful substances and physical conditions at the workplace. If the employer is found to be at fault, he or she is held liable for the full costs of injury. The worker is reimbursed for damages in the form of postinjury compensation. In theory, this approach can serve several purposes. The threat of costly lawsuits should create economic incentives that deter employers from unduly exposing their workers to hazards. At the same time, compensatory payments replace the worker's lost income and cover the costs of medical bills and rehabilitation. Like workers' compensation, legal liability is, in the first instance, a compensatory system; it seeks to make workers economically "whole" after the fact of injury. Like workers' compensation, it can also be used as an incentive system to alter the behavior of firms. Presumably, if unsafe firms are forced to pay damages to injured workers, employers will try to reduce their liability by improving working conditions.

In practice, the level of compensation, and therefore the degree to which this system creates economic incentives for prevention, replaces lost income, and takes care of other injury-related costs, depends on three related factors: the frequency with which workers sue employers; the ease with which employers can challenge workers' claims; and the legal standards applied by the courts. If this system is to function effectively, employees must recognize their injuries, be aware of their rights, and have the economic wherewithal and individual motivation to initiate legal action.

Once a case comes to court, the outcome depends on the ability of an employer to mount an effective defense. This turns on how courts and legislatures define negligence and jurors respond to competing claims. In principle, liability rests on a finding of fault based on the application of legal standards and precedents that establish the rules that employers and employees must follow if they are to avoid liability for damages. These rules can vary significantly. For example, a worker may have to prove that the employer was solely responsible for the accident or condition, and the employer may be allowed to defend himself or herself by demonstrating that reasonable measures were taken to anticipate and prevent hazards. At the other extreme, the employer can be held responsible for all accidents that occur on the firm's premises, regardless of the unsafe acts of other employees or even the precautionary measures taken by the employer. Courts can construe liability narrowly, and only replace lost income, or broadly, and compensate workers for pain and suffering. Legislatures can allow for punitive damages to increase the incentives that employers face.

Workers' Compensation

The creation of the workers' compensation system in the United States in the early twentieth century occurred as part of a more general shift toward administrative forms of conflict resolution. In keeping with this change, workers' compensation shifts the locus of decision making from the courts to boards. Although workers' compensation is an alternative to the liability suit, it, like legal remedies, approaches hazard control indirectly, through market mechanisms. In principle, it is a "no fault" system. Employers agree to establish or contribute to an insurance fund to finance compensation payments. Workers, in turn, give up the right to sue employers for damages and receive, instead, assured compensation based on a schedule of payments for various kinds of losses. Like liability suits, this approach is designed to accomplish several ends, including replacing lost income and paying medical expenses. If employers are *experience rated*, that is, charged insurance premiums based on their accident records, this system should also create financial incentives to employers to take greater care at work.

Workers' compensation systems vary cross-nationally along a number of dimensions. Some countries emphasize experience rating and use premiums as economic levers. France and Finland, for example, set employer insurance rates to maximize their deterrent effects. The Federal Republic of Germany calculates insurance premiums as a percentage of the employer's total wage costs, risk level, and accident and injury record. In contrast, the British finance their compensation program from a flat-rate premium paid jointly by employers and employees. Some countries attempt to remove obstacles to successful worker claims. Some set benefit levels generously. In Sweden, occupational diseases are compensated under the general health insurance system, which readily recognizes the impact of work on health. The Netherlands makes it possible for employees to receive full benefits regardless of the cause of the injury or disease.[10] Compared to most northern European systems, the American workers' compensation program is weak and somewhat miserly. In practice, it functions as a shared-liability rather than a no-fault system, providing partial rather than full compensation. Its deterrent effects are, consequently, limited.[11]

Factory Legislation

Factory legislation is a form of command-and-control regulation. It refers to the supervision by government of employer practices, including the setting of detailed standards that mandate changes in the design of work and machinery and that are enforced by penalty-based inspections. This legislation is a more direct form of intervention into production than negligence suits or workers' compensation. Nonetheless, it stops short of full public supervision of production. Firms remain privately owned and operated. Employers' discretion to adopt unsafe methods is limited, but government does not direct capital investment to those productive activities that are least hazardous. Nor does it promote health and safety programs, such as in-plant occupational health clinics or mandatory worker participation in plant governance, which directly encourage prevention and safer work practices.

Factory legislation is as old as industrial production. The first factory laws governed the length of the working day and workweek and the employment of women and children. Laws specifically regulating occupational health and safety appeared early in the industrial revolution. In 1802 England established regulations that required employers to ventilate and whitewash workplaces that employed more than 20 workers. Machine guards were mandated in 1844. In the 1850s and 1860s Parliament passed comprehensive statutes covering mining and factories. In 1878 the entire regulatory apparatus was consolidated under a single central authority with the power to inspect work sites. Reflecting its delayed emergence as an industrial nation, Germany began regulating work a half-century later. The first German industrial code was adopted in 1869 by the North German Federation and was extended to the newly founded German Empire in 1871. Factory inspections were made mandatory in 1878.[12]

Workplace regulation emerged in the United States on the state level. Massachusetts established the nation's first Department of Factory Inspection in 1867 and passed the first worker safety law in 1877. Other industrial states followed in the late nineteenth century. By the Progressive period, most had adopted some form of factory and mine safety legislation, most often specifying machine guarding, fire safety, and dust and gas ventilation. The federal government entered the field late, adopted characteristically indirect methods, and, for the most part, relied on the states until the passage of the OSH Act.

Today, every advanced industrial society has some form of factory legislation. The arrangements vary according to the degree of coverage, precise administrative responsibility, and penalty provisions.

But most cover their entire workforce except for self-employed or family workers. Generally, responsibility for standard setting and enforcement rests with a Ministry or Department of Labor, and the state can impose penalties on employers who fail to meet regulatory standards.

Worker Strategies Toward Work

Effective reform depends not only on state action but on how workers and unions respond to occupational hazards. As I stated in the Introduction, effective regulation requires that workers take an active role in plant governance and in national policymaking. Despite workers' interest in safe work, neither of these conditions is assured.

In capitalist democracies, workers can try to influence the conditions at work through various means, including strikes, collective bargaining, elections, and lobbying. But the structure of the political economy shapes worker demands and leads workers to consent to private control over investment. This, in turn, makes it difficult for workers to use their political rights to improve working conditions. Within the general background constraint imposed by firm profitability, employees must trade off between demands for higher wages and nonwage demands such as occupational safety and health. Since the former are compatible, in principle, with capitalist control of work and investment, and the latter are not, employees and unions are likely to be reluctant to press demands for radical changes in working conditions.

Employer strategies toward work reinforce this tendency toward economism. Elaborate job ladders, minutely divided tasks, highly routinized work, and centralized systems of labor control make it difficult for workers to coordinate efforts to take an active part in plant governance. Disorganized in this fashion, and excluded from decision making about the organization of work, employees find it difficult to learn about workplace hazards or discover the systemic roots of accidents and diseases. These factors strengthen the barriers to collective action over working conditions and encourage workers and unions to limit their demands to short-term economic gain.

Economism does not preclude collective or individual efforts to improve working conditions. Workers in a particular shop, plant, or office who feel especially threatened can strike over working conditions. If they are organized, they can negotiate improvements with employers. Individual workers can look for new jobs or use their seniority to move to less hazardous jobs. As a rule, however, economistic strategies subordinate workers' interests in occupational safety and health to more directly economic concerns, and workers' efforts are devoted to winning higher wages, shorter hours, early retirement, pensions, and the like. When issues involving working conditions are raised, economism makes it difficult for workers to press them effectively.

Alternatively, workers can adopt a more radical, political strategy and seek control over the workplace and production. This can take many forms. They can try to use their electoral and economic power to win social rights to safe work and force employers to cede some control over the organization of the workplace. They can demand publicly organized and funded occupational safety and health services, compulsory health and safety committees with decision-making power over working conditions, and worker control over these committees. They can demand works councils with codetermination rights and employer-financed health clinics. They can also press for individual rights to participate in workplace decisions, to refuse hazardous work, and to know about health and safety conditions.

In each case, however, worker demands reach beyond immediate economic benefits and job security to governance—to effective participation in the workplace and influence, through the state, over production. Seen from this perspective, conflicts over working conditions become conflicts over the organization of the political economy. Organized to make these demands, workers are more apt to be able to exercise control over workplace hazards.

A Comparative Perspective

All advanced capitalist societies rely on the three forms of state intervention outlined above. But different societies combine these forms in different ways, relying more or less exclusively on one or another element. Equally important, societies resolve conflict between employers and employees in different ways. Some rely on pluralist forms of interest representation; others are more corporatist. Labor movements vary in how they see the control of work and production. Some are more economistic than others. Finally, all these factors combine to produce relatively distinctive approaches to occupational hazards.

In the Introduction, I characterized the American approach to regulation as "liberal." Applied to the workplace, liberalism takes the following form: First, government's role in production is limited. Public authority defers to managerial control over the workplace. The private economy—firms and the labor market—are the primary mechanisms for allocating risk and protection, modified by collective bargaining agreements negotiated by employers and employees. Second, when the state does intervene, it tends to rely on factory legislation; workers are minimally involved in enforcement. Before 1970, the American approach to occupational safety and health was essentially liberal in that it rested almost exclusively on markets, private action, and a patchwork system of state laws that were poorly enforced. The OSH Act modified this system in several ways. The new law centralized existing factory legislation by establishing a single federal agency with economywide rulemaking powers. A federal inspectorate was established to monitor firm compliance and empowered to fine employers who violated standards. Worker participation in enforcement was facilitated through worker rights to participate in agency inspections.

Yet these changes deepened rather than transformed the basic approach to occupational hazards; factory legislation remained at the core of the program. The law did not require employers to establish in-plant health and safety committees that might require worker participation, or create occupational clinics to deepen the state's involvement in prevention. No attempt was made to guarantee workers the right to participate in corporate decisions about the organization of work. Finally, the act attempted to limit OSHA's powers by establishing procedural safeguards for employers' property rights, including an appeals system designed to facilitate employer challenges to agency citations.

A comparative view of workplace regulation in the United States helps to clarify how distinctive the United States approach is. The occupational safety and health programs in other Western industrial societies can be distinguished from the American approach on several dimensions.[13]

In general, few capitalist states exercise so little direct control over production. More commonly, governments either own a significant number of key industrial enterprises or use tax and credit policies to direct private investment. As a result, they enjoy a wider range of levers to influence firm policies across a range of policy issues. This gives public officials greater leverage over what employers do about working conditions.

Neocorporatist forms of interest representation are also common, in contrast to the pluralist arrangements that characterize American politics. Under a neocorporatist arrangement, social and economic policies are set by informal negotiations between organizations authorized by the state to represent the most important sectors of society. In dealing with occupational safety and health, labor unions and employer associations negotiate regulatory policy, including standards, under the auspices of the Ministry of Labor or a related body.

The resulting policies usually depend less on detailed standards. Instead, tripartite negotiations are likely to lead to compliance agreements that are flexible and tailored to the specific conditions of firms and industries. Enforcement programs, in turn, tend to rely on consultation rather than penalties.

Workers also enjoy more extensive rights to participate in enforcement programs and plant governance in many capitalist democracies. Mandatory in-plant programs and services, such as safety representatives, joint health and safety committees, and health clinics are commonplace. Codetermination institutions represent one such arrangement. *Codetermination* refers to efforts to involve workers, through their unions, in corporate decision making. In West Germany, for example, employees participate through works councils in several aspects of plant governance, and occupational safety and health is viewed as an integral part of codetermination. Workers have a legal right to participate in matters relating to occupational hazards; safety delegates must be appointed in all industrial enterprises; and statutorily mandated health and safety committees operate alongside the works councils.[14]

In societies governed by social democratic regimes, worker rights are extensive. Unions exercise significant power over national policy and workplace practices. Workers enjoy a variety of statutory rights, including the right to know about hazards and to refuse hazardous work. Health and safety committees are given a central role in plant-level decision making. In Sweden, for example, compulsory health and safety committees coordinate with in-plant union organizations and enjoy veto power over firm decisions that directly affect occupational safety and health. Many of these features are also found in Austria, Norway, and Finland.[15]

Social democratic regimes are also more likely to view occupational hazards from a public health perspective and rely less on factory legislation. The separation between the workplace and the larger socioeconomic environment is often considered artificial. This orientation can result in various kinds of programs. At a minimum, hazard-control policies emphasize health and prevention rather than punishment or income replacement. Employers, the state, or both provide occupational safety and health services as a supplement or alternative to prohibitive regulations. In-plant clinics and health services are linked with other public health programs in a comprehensive health-care network. At a maximum, employers are required to design production methods and work practices to encourage the worker's general emotional and psychological well-being, as well as accident and disease prevention. Factory-based clinics become building blocks in a preventive national health program. The Swedish program includes many of these features. The West Germans have adopted some aspects of this system.[16]

American Liberalism at Work:
A Theoretical Assessment

In drawing these comparisons I do not wish to suggest that all other advanced capitalist societies have more developed occupational safety and health programs than the United States. Many do not. On paper, worker rights are extensive in Italy and France, for example, but they are rarely enforced. Nor do I wish to suggest that the United States is unique in all regards. The German system is similar to the American one in several respects, though worker rights to participate in plant-level decisions are significantly stronger there. Nor do I wish to imply that neocorporatist arrangements are generally preferable to pluralist systems of interest representation. As I indicate in Chapter 8, neocorporatist institutions do not guarantee organized labor an effective voice in policymaking.

But this comparison does help bring the American system into focus by indicating how workers, unions, and the state can approach workplace hazards differently. Most important, it illustrates how occupational safety and health regulation reflects the assumptions of liberal reform and suggests how, in this instance, those assumptions complicate workers' efforts to win protection from workplace hazards.

First, the liberal approach does not actively involve workers in enforcement. Workers enjoy few rights to participate. Moreover, the spirit of this course reflects and reinforces economism: the state continues to respect private control over the sphere of production and hesitates to intervene in the workplace. Second, the liberal approach relies on factory legislation despite the fact that it is poorly adapted to the economic or political realities of modern industry. Production is now carried out in a large number of enterprises. There are, for example, several hundred thousand manufacturing corporations in the United States. These firms make a great variety of goods under widely varying conditions. They use complex, often new and poorly tested technologies, including recently developed chemicals. As a result, it is difficult to design regulations suited to the multiform conditions of industrial production.

Where regulations do exist, they are difficult to enforce comprehensively. The possibility of inspecting each and every firm regularly is small. The expertise required of the responsible agency is large. Furthermore, liberal-pluralist political systems often offer affected firms many opportunities to challenge standard setting and enforcement policies, often quite effectively.

Factory legislation also does little to generate worker interest in health and safety and motivate workers to take precautions. To the contrary, it encourages workers and unions to leave the development and implementation of occupational safety and health policy to central state authorities rather than become actively involved in plant-level administration. Because workers are not organized to participate in in-plant health and safety activities, public officials have to bear the entire burden of supervising working conditions. This means that the state must field a large inspectorate and rely on centrally determined standards rather than input from enterprise-level organizations. This is apt to encourage employer opposition and produce inflexible and often inappropriate rules.

Finally, factory legislation is doubly difficult to enforce when workers' organizations pursue economistic strategies. When unions do not make occupational safety and health reform a priority, they are unlikely to be able to hold public officials accountable on this issue.

In sum, both theory and comparative evidence suggest that the liberal approach to workplace regulation suffers from several institutional and political infirmities. Later, I return to the comparison between alternative policies in order to assess proposals for regulatory reform. But the next chapters bring this approach into sharper focus by considering it in detail. Chapter 2 takes a close look at the historical evolution of the American system and the factors that explain the commitment to this approach.

[2]
Before OSHA

n the years before 1970, American occupational safety and health policy strongly emphasized voluntary action by firms and employees. Despite a wide variety of state and federal programs, risk and protection were, for the most part, allocated privately. Employers were free to organize the labor process as they saw fit; while workers sought protection from occupational hazards by changing jobs or, if unionized, pressuring employers through collective bargaining and wildcat strikes. There were no worker rights to facilitate participation, or in-plant institutions to assure that participation was effective.

What state action occurred was resolutely liberal and rested on the two traditional pillars of occupational safety and health policy: workers' compensation and factory legislation. The insurance system was supposed to create incentives for employers to prevent hazards and provide injured employees with financial compensation, but it failed to function as intended. Regulatory authority was shared by national and local officials but on both levels the programs were comparatively primitive. Coverage was uneven;

standards emphasized safety rather than health and were loose and outdated; enforcement was underfunded and handicapped by jurisdictional conflicts between federal and state officials. This chapter considers the evolution of this system from the early twentieth century to the eve of the passage of the OSH Act.

Business and Working Conditions

Work has always been hazardous, but working conditions deteriorated markedly on several occasions as the American economy matured. As the factory system spread in the early twentieth century, the rates of industrial injuries and deaths appear to have quickened in mining and manufacturing, particularly in the steel mills, textile factories, and coal mines that brought modern industry to the cities and countryside.[1] Later, the petrochemical revolution introduced new kinds of health hazards into the workplace. In the 1960s a combination of economic and technical factors led to a sharp increase in the industrial accident rate.

Since the origins of modern capitalism, worker movements have resisted the worst depradations of the market and industry, and have fought to establish the rights of labor and limit the power of employers. Periodically, workers have made occupational safety and health an issue, sometimes in response to declining working conditions, sometimes as part of a wider effort to resist changes in the labor process. This issue played a major part in early efforts to build unions in mining, men's and women's clothing, and steel. The Great Depression's industrial union drives in mining, automobile manufacturing, chemicals, and rubber raised it again. Then, worker discontent over the way that management ran the shop floor fed and was reinforced by labor's efforts to win greater economic security and a larger role in governing American society. For a moment, health and safety moved toward the top of several unions' agendas and helped to stimulate industrial unionism and reorient the labor movement.[2]

But, for the most part, American firms were able to resist these efforts. As I stated earlier, the institutional arrangements of capitalist democracy facilitate business control over work and make it difficult for unions to challenge property rights successfully. Moreover, industrial interests have often taken care to anticipate and, on occasion, coopt radical movements by crafting "corporate liberal" reforms that have served to diffuse radical protest while consolidating corporate control over production.[3] Until 1970, occupational safety and health policy followed this pattern of preemptive reform. Business groups played a leading role in developing a set of public and private institutions that moderated conflict over working conditions and at the same time discouraged the development of more independent and effective programs.

Progressive Era Reforms

The first wave of workplace reform occurred during the Progressive era when manufacturing interests, concerned that the deterioration in working conditions could spark employee unrest, sought to craft public and private programs to deal with accidents and injuries. The steel industry led these efforts and used its considerable political and economic resources to organize business support for a reformed approach to work. Under the auspices of its head, Judge Gary, U.S. Steel developed the first corporate in-plant programs and joined with middle-class reformers to design the workers' compensation system. It hired safety engineers; began employee safety education; and developed the Voluntary Accident Relief Program, the first major private compensation program of its kind in the United States.[4]

The steel industry had an immediate interest in reform. Although significant sectors of the industry would later become comparatively safe places to work, steel making was difficult and dangerous work, and its hazards were well publicized by muckraking journalists and social reformers. As a result, public concern over worker health and safety focused on this industry. Equally important, steel makers were committed to keeping their factories union free and feared that accidents and injuries would stimulate worker militance.[5]

Although U.S. Steel led the safety movement, the steel industry was not alone. Its concerns reflected a widespread belief among industrialists that worker militance had to be contained through reform lest it result in effective challenges to managerial control of work or labor radicalism in politics. Other firms joined the safety movement or more broadly based reform organizations, such as the National Civic Federation (NCF). The NCF, a privately organized association of social reformers, industrialists and financiers, and moderate trade union leaders, encouraged labor–management dialogue and developed model reform legislation in areas such as workers' compensation and trade regulation. In joining this and similar groups, leading capitalists were able to play key roles in the development of private and public welfare programs.[6]

Designed to replace the existing liability system, workers' compensation was the cornerstone of the new approach to occupational safety and health. Traditionally, employers had used three common-law defenses to protect themselves from suits. The first, the *assumption-of-risk doctrine*, allowed juries to find against workers if the hazard was a known and inherent part of the job. An employer could argue that by taking the job in question, an injured employee voluntarily accepted the possibility that he or she could be injured, and absolved the employer of responsibility. The *fellow-servant doctrine*, the second of what came to be called the "unholy trinity of defenses," prevented an employee from collecting damages if the employer could prove that another worker's negligence contributed to the accident in question. Finally, the *contributory-negligence doctrine* prohibited compensation if the worker's negligence helped to cause the injury. Thus the employer who could prove that the injured employee had failed to act safely could also be absolved of responsibility, regardless of whether the firm created or contributed to the accident.

In the early twentieth century, however, several state legislatures began to liberalize the law of negligence. Concurrently, juries began to award larger settlements to injured workers and their families. Predictably, these changes generated concern among employers. As business interests often did at this time, employers turned to middle-class reformers to help them design a new approach. Working with the American Association for Labor Legislation—a research and lobbying organization consisting of liberal academics and other social reformers from the middle and upper classes—and the NCF, industrialists launched a campaign to adopt the workers' compensation system.

From the perspective of business, workers' compensation had two major advantages. First, because it was intended to be a no-fault system with fixed compensation schedules, it precluded employee suits against employers. Thus it promised to cap employers' liability and, thereby, control and regularize the costs of accidents. Second, workers' compensation took conflicts over health and safety out of the workplace and channeled them into an administrative system. There, workers confronted experts—doctors, lawyers, and public officials— rather than employers; accordingly, the issues were redefined. A power struggle between employers and employees over the intensity and conditions of work was transformed through state action into a dispute among third parties over the legal and medical principles governing worker eligibility and income-replacement schedules. In this way, the workers' attention was shifted from the organization of the labor process to the rules and regulations of a distant, often opaque bureaucratic decision-making process.

At first, some smaller firms resisted the idea, convinced that the costs of compensation would leave them at a competitive disadvantage. Represented by the National Association of Manufacturers (NAM), they opposed the plan and almost killed the movement. But larger firms agreed to pool risks with smaller firms. Once this concession was made, the system gained wide support among employers.[7]

The unions accepted workers' compensation reluctantly. Radicals in the industrial union movement argued for greater worker control of work and direct action on the shop floor. Social democratic reformers fought for strict government supervision of employer practices. More moderate in its general political orientation, the American Federation of Labor (AFL) urged that the tort system be strengthened. The AFL believed that employees would be better served by sizable cash awards under reformed legal standards than by an insurance program. But the alliance between reformist business and middle-class progressives proved decisive. Between 1913 and 1920, all but eight states passed workers' compensation laws. Belatedly, the AFL endorsed the movement.

Corporate Control over Standard Setting

However determined, standards governing worker exposure to toxic substances, machine design, and work practices are integral to most forms of state intervention. They are the major determinants of the employer's and employee's respective rights and responsibilities because they indicate what conditions government will and will not tolerate. Thus, as the state's role in this area grew, affected firms sought mechanisms that allowed them voluntarily to design and adopt occupational safety and health standards, thereby preventing mandatory regulations based on standards set by unsympathetic parties.

In some instances, in cooperation with professionals in their employ, corporations created specialized organizations and then dominated them. U.S. Steel was instrumental, for example, in the formation of the NSC in 1911 and remained a key supporter of the organization. In other cases, industry associations established for lobbying and educational purposes, such as the American Petroleum Institute, developed standard-setting activities. In still other cases, industrialists took over existing professional organizations.

The United States of America Standards Institute (USASI)—the predecessor of the American National Standards Institute (ANSI), the most important private standard-setting organization in the United States today—began in this way. The USASI was originally organized at the behest of a number of professional societies and the Departments of War, Commerce, and Navy to encourage the standardization of industrial products and processes. Then, in the 1920s, it was transformed by an influx of trade and industrial associations. In 1928 it was reorganized to reflect the dominance of corporate interests over professional societies and began to set product-safety standards that industry adopted and used to defend itself against product liability suits.[8] A long process of institution building followed, and the organization eventually branched out into workplace safety and health as an ancillary function. By the 1960s, reorganized as ANSI, it had become the principal health and safety standard-setting organization in the country.

This trend toward corporate control over private health and safety organizations was widespread. On the eve of the passage of the OSH Act, hazard identification and program development were dominated by a handful of private professional groups, including the American Society of Safety Engineers (ASSE), the Council on Occupational Health (COH), the American Hygiene Association (AHA), and the Industrial Hygiene Foundation (IHF). For the most part, these organizations deferred to management, tied to employers by financial considerations, membership, and organizational affiliations.[9]

The NSC stood at the center of this private network. Over time, it had gained considerable legitimacy as a disinterested party, despite the instrumental role that U.S. Steel had played in its formation. Nonetheless, it remained a captive of its member firms: three-quarters of the NSC's trustees were corporate managers, and it functioned as a public relations agency and corporate think tank rather than an independent research body. Because it represented the larger, "safety-conscious" firms, it tended to be more liberal than other organizations in the field.[10] But, like the rest, the NSC developed and promoted preventive strategies that coincided with corporate control of production, personnel relations, and plant operations.

The ANSI supplemented the NSC by developing what were called "consensus standards." These were guidelines that employers could, if they wished, apply to their practices. Like the NSC, ANSI claimed to represent a wide constituency, and it regularly invited representatives from business, labor, consumer groups, professional associations, and government agencies to take part in its activities. The standards committees were, in turn, made up of experts from a broad range of these interested organizations. As a result, ANSI was accorded special status by federal and state agencies.

In practice, ANSI, like the NSC, was dominated by the companies and trade associations that helped to organize and finance it. Members enjoyed voting rights on policy issues and dominated the standards committees. As of the late 1960s, only 6 trade unions were authorized to participate in ANSI committee deliberations, compared to 160 trade associations. In the vast majority of cases, government, union, and consumer representatives combined constituted a minority of committee members.[11] Scientists employed in the private sector supplied the technical expertise. Proprietary groups (i.e., industry associations) provided summaries of company practices and in-house standards that were used as the starting point for committee work. Proposals by ANSI were also reviewed by industry representatives and a Board of Standards Review. The board was dominated by industry representatives and enjoyed plenary power over all standard-setting activities.[12] As a labor department report concluded:

> One of the weaknesses of the standards process, in respect
> to occupational safety and health standards and

consumer goods standards has always been that the
consumer, the working man or the housewife, has
always spoken with a very weak voice in the councils of
the standardizing bodies.[13]

Nearly all of these private and professional organizations sub-
scribed to and propagated a set of principles that legitimated corporate control of work and the privatization of the workplace. According
to this view, most accidents were caused by what they called "worker
error" or "unsafe acts." Employers, in contrast, were viewed as en-
lightened, albeit self-interested, guardians of their workers' welfare.
Firms, especially large firms, were careful to check new substances
before they entered the workplace and take the precautions necessary to assure that work was safe. Indeed, the employer's self-interest
served the worker: productivity losses and high insurance premiums
resulting from accidents, deaths, and diseases led the rational firm to
make work as safe as possible. Smaller businesses might not take
equal care, of course. They lacked the money and expertise to match
their larger competitors. But their mistakes did not justify wholesale
state intervention. Public subsidies for education and training could
bring the smaller firms' programs up to par.[14]

In fact, private action was held to be more efficient than government regulation under almost all circumstances. Government agencies could make existing industry and professional codes available
to all firms, conduct research in industrial hygiene, and hold conven-
tions and meetings to encourage interest in worker safety and health.
In contrast, detailed codes were unlikely to prevent accidents; indeed,
they could have little effect on prevention, since the vast majority of
accidents were caused by the employees themselves. Actually, government regulation was likely to discourage accident prevention. It
led firms to concentrate their efforts on compliance with artificial
codes rather than devote their time and money to the more important
effort of educating workers. Thus, private organizations should continue to have the responsibility for developing safety and health standards, and employers should have the freedom to apply them flexibly.

Only the American Conference of Governmental Industrial Hygienists (ACGIH), representing professionals who worked for the government, could claim any real independence from the corporate

sector. The ACGIH was created in the 1930s by reformers in the industrial hygiene profession who were dissatisfied with the biases of the corporate-dominated American Industrial Hygiene Association. Alone among organizations in this network, ACGIH promoted awareness of occupational health problems. It set the voluntary threshold limit values (TLVs) for hundreds of toxic substances that later became the basis for many of OSHA's health standards. As a rule, however, corporations did not adopt these standards; and most government agencies deferred to this industry practice.

In-Plant Programs

At the workplace, in-plant programs were limited in scope and purpose. The evidence does not allow for a precise calculation of the extent of corporate activities before 1970, but one reliable estimate of current corporate efforts indicates that most corporations do not provide permanent medical facilities. Only the largest establishments provide any full-time staff; this usually means a single occupational nurse. Almost all work sites with less than 500 employees—approximately four-fifths of the nation's workers work in these establishments—lack any organized preventive program. It is safe to assume that in-plant programs were even less developed prior to 1970.[15]

Even where corporations mounted permanent in-plant programs, these efforts were subordinated to management personnel policies that often conflicted with sound occupational safety and health policies. Health and safety directors were below plant and personnel managers in the corporate hierarchy. In-plant programs were not designed to anticipate and prevent disease and injury; they were organized to screen employees who might be unfit for work. They stressed the physical, mental, and social adaptation of the worker to the workplace, rather than the reverse. Safety, not health, was the priority. Injuries were treated on an emergency basis.[16]

In this context, prevention devolved into worker education; worker education devolved into propaganda extolling the virtues of management programs and blaming worker carelessness for accidents and injuries. Health and safety committees, where they existed, were usually shut out of corporate decision making about plant-level policies that involved significant expenditures or changes in work relations. One unionist told Congress of his experience with monthly committee meetings in this way:

> The committee is very sincere, the safety committee. They
> devote much of their time to making the plant safe.
> Management are very good listeners, and they take minutes of
> our talks. When it comes to getting minor stuff fixed we have
> no problem but when it comes to something big, where the
> costs are a lot of money, then we get into a lot of difficulties.[17]

Organized Labor and the Workplace

Although business strategies were the driving force behind corporate dominance of health and safety at work, the unions helped to reproduce it by pursuing an economistic strategy that ceded control of the labor process to employers. In the late nineteenth century, Samuel Gompers, first president of the AFL, committed the newly formed federation to a "voluntarist" approach to industry and government: workers struck to build and protect unions and voted to win moderate reforms. Despite major changes in American capitalism and the labor movement between the two world wars, the labor movement's strategy remained remarkably consistent in this area. After World War II, the AFL-CIO stressed organizational security and economic gains for workers rather than participation at work or public control over industry.

The Postwar Accord

Rank-and-file workers did challenge corporate decisions about health and safety, and struggled to improve working conditions, particularly through local actions, after the war. But these struggles were framed by a general labor-management "accord" that helped to reinforce the depoliticization of work and private control over health and safety organizations.

By *accord*, I refer to the institutional arrangements that organized industrial relations in the United States from the New Deal through the

1970s. These arrangements evolved incrementally during the 1930s and 1940s in response to the labor unrest of the Great Depression and World War II. Finally consolidated in the mid 1950s, the accord is best understood as an implicit social contract, or "truce," between American workers, represented by the AFL and CIO unions and the larger multinational corporations.[18]

Intended to secure labor peace, the accord established a routinized system of political and economic conflict with a fairly clear set of rules. These rules were institutionalized and codified in laws regulating collective bargaining (the Wagner, Taft-Hartley, and Landrum-Griffin acts), social security reforms, statutes and organizations that regulated economic growth (the Council of Economic Advisers, the Employment Act), and employer-employee bargaining strategies.

Characteristically, workers accepted managerial control over production and capitalist control over investment. They also limited their political activities to demands for economic growth and social security rather than the transfer of income and wealth between classes or the reorganization of the labor process. In return, corporations accepted a greater role for unions in national politics and moderated many of the most extreme forms of labor control. They reduced the arbitrariness of plant-level supervision and developed new personnel programs and managerial practices that stressed conciliation and arbitration.

These arrangements did not preclude using state power to advance worker interests at work. The union movement lobbied Congress to strengthen the laws governing collective bargaining so that employees might have greater leverage on the shop floor. Workers retained the right to strike over the conditions of work. But the accord shifted the focus of worker activity away from struggling with management over the organization of production to lobbying public officials for policies promoting full employment and economic security. At the same time, many unions negotiated contracts that consigned conflicts over plant administration to local union bargaining or a highly bureaucratized grievance procedure. In combination, these institutions and practices reinforced corporate control of work and undermined efforts to improve occupational safety and health.

In several respects, the accord was a rational response by organized labor to the subordinate position of American workers in the political economy—a strategic concession designed to maximize workers' limited power resources. The labor movement was weak. Only a small proportion of the labor force was organized; those who were in unions lacked central organization. There was no distinctive labor party to focus worker political activity, and despite the mass protest of the New Deal period, workers remained demobilized in their communities and at the workplace. Moreover, as anticommunism waxed and labor militance waned in the late 1940s, employers used their political power to win labor statutes that banned secondary boycotts and discouraged other cross-sectoral forms of coordination by workers. These laws then reinforced demobilization, disorganization, and divisions among workers.[19]

Once the arrangements were established, they created incentives for unions to continue along these lines. Barring a classwide political mobilization—unprecedented and unlikely in the United States—opposition to the demands of business for unilateral control over work, investment, and technology were likely to be costly and unsuccessful. In contrast, by granting management these prerogatives at the workplace, organized labor was able to win substantial organizational and economic gains that had previously been denied to unions and workers, including legal protection to organize, substantial wage gains, and a public commitment to high levels of employment.

Intraunion political and organizational factors also helped channel worker activity in these directions. The labor leaders of the 1950s and early 1960s were the survivors of labor's civil war between left and right. Most were firmly procapitalist and hostile to the more radical visions of industrial democracy and worker participation that had flourished in the 1930s. And the union bureaucracies benefited as organizations from routinized collective bargaining and labor peace. Agreements to check off union dues from workers' paychecks maintained union treasuries with a minimum of effort. Favorable National Labor Relations Board decisions protected established union fiefdoms.

The Accord and Health and Safety

However rational it seemed on its own terms, the accord ultimately proved debilitating because it reproduced rather than reduced those factors that constrained worker activity and limited workers' power as a class. The unions' single-minded pursuit of short-term material gain undermined the militant spirit that had helped to rebuild the labor movement in the 1930s. The union–Democratic party alliance encouraged workers to depend on presidential leadership rather than their own organizations for social reform. As amended by the Taft–Hartley Act in 1947, the National Labor–Management Relations Act precluded the kind of classwide organizations that made it possible to surmount the institutional obstacles that workers faced.

As for health and safety, the industrial relations system as a whole discouraged attention to the substantive features of shop-floor relations, including all but the most immediate working conditions. In collective bargaining and in politics, the hazards of work became a secondary issue. Conflicts over working conditions were routinely channeled into the administrative apparatus established to handle local disputes between managers and employees. Occasionally, workers struck over working conditions, but the unions emphasized other issues. Like economism in general, this approach to worker health and safety was rational on its own terms. Divisions within the union movement discouraged radical action in the name of health and safety. The former AFL craft unions were generally indifferent to worker control and job safety. The construction and building trades were difficult and dangerous jobs, but the unions in these occupations preferred to negotiate for risk premiums rather than challenge employer prerogatives.

Moreover, all unions faced disincentives to challenging workplace practices. Conflicts with employers over the conditions of work were costly and did not yield tangible gains. Indeed, the record was bleak. Only the United Auto Workers (UAW) was able to win a major contractual concession on health and safety before the passage of the OSH Act.[20] Many union leaders believed that employers would abandon plants and factories rather than increase investment in health and safety. The Papermakers Union in New Jersey is a tragic example. Fearful of the prospects of job loss, it was reluctant to challenge Johns Manville over the hazards of asbestos despite long-standing evidence of the problem. The Mineworkers and Textileworkers had similar concerns and were similarly passive in the face of life-threatening conditions.

The laws governing collective bargaining also discouraged union struggles over working conditions. Administrative and judicial decisions limited the scope of permissible worker activity at work. Prior to 1966, health and safety was not a "mandatory" subject of collective bargaining; managers were bound to discuss it but they could refuse to compromise on it. If unions pressed the issue to the point of impasse, however, they could be penalized for bad-faith bargaining.

Some union leaders actually discouraged worker demands for health and safety because they associated it with rank-and-file challenges to their leadership. Given the nature of the issue, and the workers who remained interested in it, the two were often linked. Dissident Teamsters, for example, made truck safety a major demand, and the Teamster leadership rejected it. The rank-and-file campaign for safety legislation and black lung compensation in the late 1960s fed opposition to Tony Boyle's leadership of the UMW, and Boyle initially tried to block the movement. Boyle endorsed demands for federal legislation only after his opposition to health and safety reform threatened his hold over the union.

For all these reasons, occupational safety and health dropped off organized labor's postwar political and economic agenda. Only the UAW tried to mount an independent program. It had safety and occupational disease services and employed three full-time staff people. Most unions did little to prepare themselves or their workers to recognize or remedy hazards, let alone challenge existing corporate arrangements. Some actively participated with management in the private health and safety network. Union health and safety directors from the Operating Engineers, the United Steelworkers, the Papermakers and Paperworkers, and the United Rubber Workers participated in NSC programs. They won awards and ran articles in their newspapers about safety practices. A few AFL–CIO representatives sat as token members on ANSI's standard-setting committees. For the most part, however, the labor movement was indifferent to the problem. Unions bargained away safety and job control for productivity-based wage gains, health programs, pension plans, and unemployment insurance.

Collective bargaining agreements reveal in quantitative terms the labor movement's position on health and safety. A Bureau of Labor Statistics (BLS) study indicates that before 1971, the typical contract contained only rudimentary protections (see Table 2.1). Less than one-fourth stipulated employee or union rights to information, safety

Table 2.1. Collective Bargaining Provisions Before 1971

Provision	Percentage of agreements on health and safety that refer to selected provision, by frequency of reference
Accident procedures or compensation	63.0
Safety equipment	55.0
Sanitation provisions	51.9
Employer pledges of compliance with law	41.4
General policy	39.8
Employee rights with regard to safety	23.5
Safety committees	22.9
Physical examinations	21.8
Union-management cooperation pledges	20.6
Joint safety committees	19.5
Discipline for noncompliance	18.5
Union rights with regard to safety	15.8
Safety inspections	13.2

Source: Adapted from Winston Tillery, "Safety and Health Provisions Before and After OSHA." *Monthly Labor Review* 98, no. 9 (1975), table 2, p. 42.

inspections, safety committees, or physical examinations. Moreover, when there were provisions, they had little force. Safety committees were advisory; employee rights to refuse unsafe or hazardous work, when mentioned, were left to further negotiation.

Public Policy Before the OSH Act

Prior to OSHA, occupational safety and health policy simply ratified the subordinate position of workers at work and reflected employers' interests in labor control and a depoliticized workplace. Both the

workers' compensation system and federal and state regulatory efforts supplemented rather than supplanted corporate programs. Both
were extensive, but neither effectively challenged managerial control
of work or significantly altered the incentives that employers faced
when making health and safety decisions.

Workers' Compensation

Given the options in the early twentieth century, the workers' compensation system was a major advance over the late-nineteenth-century
approach. The courts interpreted the "unholy trinity" of employer defenses liberally, and this all but precluded successful employee suits.
As a result, tort remedies proved inadequate. One contemporary
observer estimated that only 15% of injured employees recovered
damages.[21] A reformed liability system might have served workers
better, but, as the AFL learned, organized labor was not in a position
to change the law of negligence over the concerted opposition of
employers and middle-class social reformers.

After the Progressive era, workers' compensation grew to include
50 state programs regulated by industrial boards and commissions.
Although the details of the programs varied somewhat, all followed
a similar pattern. They established a no-fault insurance program
in which employers carried liability insurance to cover workplace
accidents and disease; employees were compensated for wage loss
and medical costs according to schedules and rates established by
state legislatures. In return for compensation, workers gave up their
right to sue employers for nearly all accidents and injuries.

In this way, the workers' compensation system attempted to achieve
the goals of income security and prevention within a market framework rather than provide for the direct regulation of work by public
officials. Compensation was intended to replace lost wages due to
accidents and cover workers' medical expenses. The system was also
designed to send market signals to employers by linking insurance
premiums to employers' accident rates. Employers could then make
economic decisions about how much safety they wished to provide
and in what ways: safer employers would pay lower premiums;
employers who failed to provide safe workplaces would choose, in

effect, to devote a larger percentage of their expenditures to workers' compensation insurance.

As it developed, however, the system proved severely flawed, as studies done in the 1960s and early 1970s indicate. To begin with, workers' compensation failed to cover all employees; 20% of workers were totally unprotected, and they were concentrated in the lowest-wage occupations; only 12 states covered farm workers; only 8 states covered domestic workers.[22] In addition, covered workers were rarely paid for their full losses. Compensation was uneven and lagged behind wages and the cost of living. Short-term injuries were under-compensated because of waiting periods for filing claims. Workers suffering from permanent total disability were especially hard hit. Statutory limits on maximum benefits and on the total number of payments allowed limited their awards. The maximum weekly benefit for all states averaged less than one-half of average weekly wages; actual payments were lower.[23] In 42 states, maximum benefit levels did not meet the 1966 standards for a poverty income.[24]

The system also proved nearly impermeable to worker claims for compensation for occupational disease. Some states limited coverage to lists of specified diseases, and coverage was generally outdated and overly restrictive. Strict time limits for filing claims (often less than a year) were regularly applied. Difficulties in assessing causation, long latency periods, worker ignorance of health hazards, and employer resistance to disease-related compensation combined to depress the number of claims. Research indicates that, at the most, 3% of occupational disease cases actually resulted in compensation.[25] Only a minority of employers paid experience-rated insurance premiums. In sum, employers were not forced to internalize the true economic costs of accidents and injuries. Whatever its stated goals, the system failed to create sufficient economic incentives to prevent accidents and diseases.

State Programs

Workers' compensation was supplemented by state regulatory programs. Local regulation began early in the industrial revolution, and most states had some form of safety legislation by 1900. In the 1880s state associations of labor commissioners and factory inspectors were founded to share information and coordinate efforts among the various states. In 1914 this movement resulted in the establishment of the International Association of Governmental Labor Organizations (IAGLO) to coordinate efforts among federal, state, and provincial labor officials in the United States and Canada. By 1970, approximately 90% of all workers in the United States were covered in some way by state regulatory laws. These agencies, in turn, enjoyed a full range of regulatory powers from inspection to penalties, covering both chemical and physical hazards.

But wherever they emerged, these programs quickly adapted to corporate control of work. They deferred to voluntary corporate safety efforts, the insurance carriers who wrote workers' compensation, and private safety organizations such as the NSC. As a result, nearly all programs were weak and ineffectual. Almost all of them focused on safety and virtually ignored health hazards. Less than one-third had regulations covering biological hazards.[26] Industrial hygiene programs were, according to a Department of Labor study done in preparation for the debate over OSHA, "with but few possible exceptions . . . totally inadequate."[27]

Most state programs were also starved for resources. Thirty-eight states responded to a labor department survey of fiscal 1968 commitments. On average, they spent 48 cents per nonagricultural worker on occupational health and safety regulation.[28] Staffing under these financial constraints was predictably small. The most comprehensive survey of field activities reported a total of 1803 safety inspectors and 489 occupational health and industrial hygienists in all 50 states in 1969. Less than two-thirds of the safety inspectors were assigned exclusively to general safety. Nearly a fifth inspected boilers, elevators, and fire hazards. Only a handful of inspectors specialized in industrial disease.[29]

State efforts were also hampered by legal and administrative problems. A majority of states divided authority for workplace protection among at least three separate agencies, including industrial commissions and labor and health departments. Most states also restricted agency enforcement powers significantly. Twenty-one states did not allow inspectors to shut down machinery in imminent danger situations; 16 states had no criminal sanctions against deliberate violations of the law; 5 states did not give inspectors the legal right to enter premises without the permission of the employer. Most states relied on weak and outdated standards.[30]

The emphasis on local rather than national authority further undercut protection by leading to uneven enforcement. The variation in state efforts was extreme and for the most part corresponded to variations in industrial development and the strength of the labor movement. California, Illinois, Michigan, New York, Massachusetts, and Pennsylvania mounted significant regulatory efforts.[31] New York, California, and Pennsylvania accounted for nearly one-half of the nation's total inspectorate. The Mountain and southern states spent the least per worker on health and safety; northeastern states spent considerably more. Oregon, Washington, and Colorado—each the home of a particularly dangerous primary or extractive industry—had the best-financed programs.[32]

Even states with relatively sound programs depended on voluntary compliance by industry rather than enforcement by state agencies. They devoted their resources to education and training and tried to minimize adversarial relationships. States often warned employers in advance of inspections and deemphasized penalties for violations. A study of New York's record of inspections during 1968 found that prosecutions were rare. Of the more than 10,000 violations discovered by inspectors, only 442 cases were referred to prosecution. Only 6 cases resulted in fines.[33]

Federal Programs

Federal authority was equally compromised. Historically, Washington had deferred to the states and allowed them to set the regulatory agenda. Although the first federal laws and programs were developed during the Progressive era, the federal effort was minimal prior to the Great Depression. In the 1930s federal authority was augmented through the Walsh–Healey Public Contracts Act, passed in 1936 as a stopgap measure to set minimum wages on government contracts until the constitutional issues raised by a general minimum-wage law were resolved. It regulated working conditions to assure that firms competing for federal contracts did not cover increased wage costs by cutting back on worker health and safety. Though not central to the act, this provision resulted in extensive federal jurisdiction over working conditions as federal contracting increased during and after World War II, particularly in defense-related sectors.

Public concern over the conservation of labor during World War II also stimulated interest in health and safety. The social security system was used to finance state programs through grants-in-aid to state agencies, and war boards urged employers to improve working conditions. Once the war-induced labor shortages disappeared, however, the federal government retreated from its wartime efforts. A series of Presidential Conferences on Occupational Safety was inaugurated in 1946, but these biannual meetings turned into forums for self-congratulatory speeches by corporate managers and safety professionals. The private organizations regained their customary prominence in the field: in 1953 Congress chartered the NSC as a nongovernmental public service organization; the Department of Labor (DOL), in turn, adopted privately developed standards to implement its Walsh–Healey jurisdiction.

Without strong support from organized labor, the DOL could or would not seek congressional support for an expanded health and safety program. The department's funding requests were perfunctory and unimaginative, and it acquiesced to state governments' demands for local control of workplace regulation. From the late 1940s to the mid 1960s, the DOL lobbied only for grants-in-aid to the states. Even these bills were routinely defeated, often without hearings or any serious lobbying by the secretary of labor or the states that would have benefited.[34]

Department of Labor jurisdiction expanded in some areas. In 1958 Congress amended the Longshoremen's and Harbor Workers' Compensation Act to allow the labor department to operate a safety program on the docks and in shipbuilding and repair. In the mid 1960s the National Foundation on the Arts and the Humanities Act and the Service Contracts Act extended coverage to groups, agencies, and individuals receiving grants or providing services. By the late 1960s, the department's authority covered over half of the nation's workforce (see Table 2.2).[35]

Unfortunately, like the states, the department's resources were woefully inadequate. In the late 1960s the Labor Standards Bureau (LSB)

Table 2.2. Department of Labor Jurisdiction, 1967

Authority	Workers Covered (labor force = 75 million)	Expenditure per Worker[a]
Maritime Safety Act		
Longshoring	103,000	$8.84
Shipyards	120,000	3.79
Marine construction	20,000–100,000	0.0
Walsh–Healey Act	25,000,000	0.01
Service Contracts Act	6,000,000	0.06
Vocational Rehabilitation	150,000	0.0
Arts and Humanities Act	10,000	0.0
Federal Labor Standards Act	8,250,000	0.01
Federal Employment Compensation Act	2,800,000	0.07
TOTAL	42,493,000 (56.7% of labor force)	0.05 (average)

[a]The total expenditure per program divided by the number of workers covered by that program.
Source: Current Department of Labor Responsibilities and Activities, Attachment Six (Washington, D.C.: Department of Labor, 1968).

spent less than $3 million on all its safety activities combined. Half of this effort was devoted to maritime regulation; 23 people implemented the Walsh–Healey and Public Service Contract acts with a budget of less than $400,000.[36] Not surprisingly, fewer than 5% of establishments covered under the terms of the act were inspected in any given year.[37]

Instead, the federal government relied on the underfunded and understaffed state agencies for enforcement. But without federal assistance or encouragement from organized labor, state efforts also atrophied. From 1950 to 1969, state agencies lost personnel and programs in occupational health despite a doubling of the states' efforts in the general area of health services. Sadly, occupational health programs were often raided to staff and fund newer programs in air-pollution and radiation control.

Jurisdictional disputes compounded these problems by hindering federal-state cooperation. According to a 1968 report to the Department of Labor, there was "an almost complete lack of meaningful interaction between Federal and State officials on basic policy, program and administrative issues."[38] The DOL tried to coordinate state-level activities by chairing the International Association of Industrial Accident Boards and Commissions (IAIABC) and IAGLO. It attempted to influence the states through the grants-in-aid programs. The LSB drew up model workers' compensation laws and encouraged states to upgrade their standard-setting and enforcement activities. But these efforts were ineffectual. The states jealously guarded their prerogatives and rejected federal funds that promoted federal supervision of local efforts.[39]

In the end, like the states, the federal government was forced to depend on the voluntary compliance of private firms. Therefore, like the states, the DOL took care to encourage employers' cooperation. According to one audit, only 34 formal complaints were issued, and only 2 firms were penalized in fiscal year 1969, despite the fact that 95% of the establishments inspected that year were in violation of DOL standards.[40] Instead of citing violations, the department emphasized consultation and training for state and private safety organizations. Instead of setting standards independently, it cooperated as an interested organization with private standard setters and adopted ANSI standards to implement the Walsh–Healey program.

The Public Health Service (PHS) was authorized to engage in research on worker health problems, but it was even more passive than the DOL. The Bureau of Occupational Safety and Health (BOSH) had a minuscule budget; in 1955 its total allocation for occupational safety and health amounted to less than one cent per worker per year.[41] In combination, BOSH and the National Institute of Environmental Health Sciences spent less than $9 million on environmental health in fiscal year 1970. Equally important, the PHS's leadership respected private control of worker health and safety. Not only did the PHS eschew enforcement, but its research efforts deferred to industry professionals, the American Medical Association (AMA), and local authorities. Despite mounting evidence here and abroad of the link between workplace exposures and disease, it failed to investigate asbestosis, byssinosis, or coal miners' pneumoconeosis. The PHS actually considered abolishing BOSH altogether.[42]

The Health and Safety Crisis

Given the tendencies inherent in corporate control of the labor process, public deference to employer decision making left workers vulnerable to market processes and management decisions that emphasized production and profitability over hazard control. As a result, the business cycle and the advance of technology determined the quality of working conditions in the postwar period.

The consequences for workers varied by firm size and economic conditions. In manufacturing, mid-size firms employing between 50 and 500 workers were the least safe; establishments with fewer than 20 workers or more than 1000 employees were the safest.[43] Accident rates were also sensitive to changes in the overall economy, declining during the years of sluggish growth from the late 1940s to the early 1960s, then rising dramatically in the boom years of the 1960s.[44] Trends in worker exposure to health hazards cannot be measured as easily. But what we know about occupational diseases suggests that workers remained vulnerable here as well. Working conditions in the dusty trades, such as textiles and mining, remained dangerous, whereas changes in industrial technology exposed a growing number of workers to the hazardous by-products of the petrochemical revolution and the stress of work in increasingly bureaucratized workplaces.

Job Safety

Changes in injury rates seem to have responded to several forces. As a rule, they are extremely sensitive to the business cycle. The reasons for this are straightforward. As business expands, employers hire new workers, open new plants, bring new machinery into use, and push

Before OSHA

Figure 2.1.   Manufacturing Work Injury Rates, 1956–1970

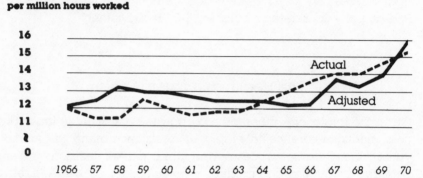

**Lost-time injuries
per million hours worked**

Note: Injury rates are adjusted for cylical changes in overtime, hiring, and capacity
utilization.
Source: Robert S. Smith, *The Occupational Safety and Health Act* (Washington, D.C.:
American Enterprise Institute, 1976), figure 1.

existing facilities to their limits by running operations around the
clock and speeding up assembly lines. Each of these factors tends to
increase the likelihood that accidents will occur and workers will be
injured. New workers are apt to be young and inexperienced, and
these groups tend to have higher injury rates than older, more experienced workers. Both management and workers are likely to be
unfamiliar with new machinery and err in its use. Pressing existing
facilities and production lines to their limits probably compounds
these problems. In combination these factors make it likely that injury
rates will rise as the economy heats up. In business downturns, all of
these factors are reversed, and injury rates tend to decline.[45]

Robert Smith's analysis of injury rates in manufacturing indicates
that, as this model predicts, cyclical factors were at work in the postwar trends (see Figure 2.1). Nevertheless, as Smith notes, business-
cycle factors alone do not explain the deterioration in accident rates
after 1958. While an adjustment for cyclical influences affects how we
date the upward trend—Smith's adjusted data move it back from
1963 to 1966—even these corrected data indicate a sharp rise in
injury rates in the late 1960s.[46]

Employer strategies in the face of declining profits, and the impact
of the Vietnam war mobilization on production, appear to account for

Figure 2.2. After-tax Corporate Profit Rate, Manufacturing, 1955–1970

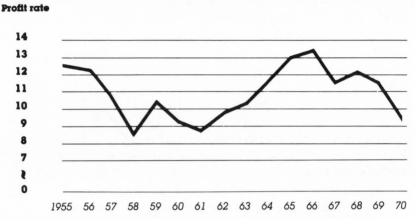

Note: "Profit rate" is the ratio of profits after income taxes to stockholders' equity.
Source: *Economic Report of the President, 1985* (Washington, D.C.: GPO, 1985), table B-86.

the remaining unexplained increase. There is some dispute about the precise role of employer strategies vis-à-vis work. Michel Aglietta suggests that working conditions deteriorated in response to the "hardening of the class struggle" between 1958 and 1966. According to Aglietta, management intensified work in the 1950s in order to improve sluggish profit rates. Unions, weakened at the workplace by the postwar accord, were unable to resist employer efforts and speed-ups. Declining job-safety standards resulted.[47] Bowles, Gordon, and Weisskopf concur. But they date the management offensive to the mid 1960s after profit rates had declined dramatically (see Figure 2.2).[48] Smith's corrected data appear to support this latter view: accident rates rose after profit rates declined. These interpretations are not mutually exclusive. Profits rates were sluggish in manufacturing in the late 1950s. Employer efforts to intensify work may have begun then and accelerated in the face of continuing economic problems.

Finally the Vietnam war also appears to have contributed to the deterioration in injury rates. The war induced an exceptional economic boom, pushing American industry to near-full capacity. This economic mobilization may well have undercut worker safety. In fact, accident rates climbed sharply in industries most directly affected by wartime production, including primary metals, fabricated metals, and machinery; the rate of increase doubled in ordnance industries. At the same time, injury rates declined in contract construction between 1958 and 1968 and remained stable in mining over the same period.[49]

The Health Crisis

The nature of occupational disease and the extent of worker exposure to it is more difficult to assess than all accident and injury rates. The precise causal links between exposure to toxic substances and physical agents are often unclear. Many diseases are caused by multiple factors that interact in complex ways; the hazards of workplace exposures can be exacerbated by general environmental conditions and family histories. Workers who handle asbestos and smoke, for example, are more apt to contract lung disease than those who do not smoke. It is likely that living in a polluted neighborhood or drinking contaminated water increases a worker's susceptibility to developing work-related cancers. To compound the problem, the long latency periods of diseases such as cancer make it difficult to identify causal links.

Corporate control of research in this period further frustrated efforts to pin down these relationships. Several studies were carried out under the auspices of individual firms and trade associations concentrated in particularly hazardous industries. Du Pont, Johns Manville, the Manufacturing Chemical Association, and the American Textile Manufacturers Institute investigated health hazards in their plants. But these studies were often designed or presented to respond to private suits and possible government intervention rather than provide independent measures of the hazards at work. Checked against workmens' compensation data and in-depth state studies, large discrepancies routinely appeared between what was voluntarily described and what was actually occurring.[50]

Despite statutory commitments to conduct research efforts, neither the Department of Health, Education, and Welfare (HEW) nor the Department of Labor provided independent data. The Public Health Service did not have legal authorization to enter workplaces and was totally dependent on company reports. The LSB had the authority to inspect workplaces to enforce Walsh-Healey standards but did not take advantage of its jurisdiction and was also dependent on data voluntarily supplied by firms.

A few journalists, doctors, and labor activists attempted to fill this void. Frank Wallick, editor of the UAW's staff newsletter, wrote trenchant pieces informing the union's membership of the hazards they faced. Paul Brodeur published a series of articles in the *New Yorker* on asbestosis at a Pittsburgh-Corning plant in Tyler, Texas. Dr. Irving J. Selikoff, director of the Environmental Sciences Laboratory at Mount Sinai School of Medicine in New York City, mounted a major research effort that finally demonstrated the link between exposure to asbestos and pulmonary asbestosis among insulation workers. Ray Davidson, editor of the Oil, Chemical and Atomic Workers (OCAW) newspaper, published a compelling account of health hazards in the oil and chemical industries.[51]

For the most part, however, these studies were handicapped by industry's near monopoly on information about occupational health and safety. Only firms in hazardous industries had unlimited access to the reports of company doctors, as well as detailed information on the kinds of substances routinely handled at work. Without this information, health experts had difficulty establishing the links between workplace exposure and occupational diseases. Personal physicians knew little about working conditions and saw too few employees from a single plant to suspect or establish an association between work and illness. Health scientists had to rely on incomplete data taken from death registries and personal employment histories. Some corporations hid evidence of health hazards for fear that publicity would lead to lawsuits and a flood of workers' compensation claims.[52]

Given all these factors, it is understandable that precise figures on the scope of occupational diseases and deaths varied considerably; they still remain open to dispute. The most commonly quoted estimates suggest that 390,000 cases of occupational illness occur each year and that as many as 100,000 workers die from workplace health hazards. The figure of 390,000 illnesses was cited during the congressional debates on the OSH Act and repeated in a 1972 HEW study of worker health. The estimate of 100,000 deaths is extrapolated from several early epidemiological studies and an analysis of a 1951

British death registry. Neither estimate can be quoted with tremen-
dous confidence; other studies offer widely ranging estimates of between 10,000 and 210,000 worker deaths per year.[53] Nonetheless, it is
reasonable to conclude that, by the 1960s, work threatened workers'
health as much, if not more, than their safety.

The Policy Choice

By the mid 1960s, a growing number of public officials and health professionals recognized that the existing system for protecting workers
from occupational hazards was breaking down. Taking a lesson from
the nascent environmental and public-interest movements, they and
a small group of labor activists began to promote the idea of workplace reform. Their efforts were initially resisted by employers and
many of the professionals employed by the private health and safety
organizations. The existing system had survived more or less intact for
a half century, and corporate managers were intent on retaining
control over work and the labor process. But the political climate was
receptive to social change, and workplace safety and health reform
was caught up in the rising tide for social regulation of industry and
the marketplace.

Once the reformers succeeded in placing the issue on the policy
agenda, public officials had to choose among the particular forms of
state intervention described in the previous chapter. Reform could
have strengthened the liability laws and facilitated worker suits.
Legislative reform could have made it easier to prove employer negligence and secure punitive damages. Or workers' compensation
could have been reformed: the system could have been nationalized;
uniform standards could have been applied; coverage could have
been extended to all workers; all firms could have been experience
rated; and disease-related claims could have been facilitated. The
entire emphasis could have been shifted in keeping with the new
attention to the total environment: the workplace could have been
integrated into a reformed public health program, and resources
could have been devoted to creating in-plant clinics linked to a
national health-care system.

Reformers also faced the issue of worker participation. Factory legislation can facilitate active participation by employees in the determination of working conditions and the operation of in-plant programs. Or it can rely on decisions by professionals and bureaucrats. Similarly, standard setting can be depoliticized and lodged in an independent commission or subject to political negotiations between labor and capital.

In the end, the Occupational Safety and Health Act strengthened the existing regulatory apparatus and reinforced rather than substantially reformed the liberal mode of regulation already in place. The existing liability laws and workers' compensation system were left unreformed; the public health option was considered and dropped. Some steps were taken to facilitate worker participation, but these were small. The next chapter considers the factors that made occupational safety and health a policy issue in the 1960s and that led Congress and the White House to this choice.

[3]
The Origins of the OSH Act

The OSH Act, on one level, dramatically restructures the relations between employers and employees. It creates new rights to health and safety and empowers the federal government to enforce them. On another level, however, the new law is conventional. In form, the program builds on and reflects the assumptions of the old system. The state continues to rely on a few narrowly defined kinds of state action. The agency uses administrative rulemaking to set standards, and rulemaking is based on the expertise of professionals. Enforcement depends on penalty-based inspections rather than workers; the workplace remains depoliticized.

This chapter considers the factors that led to the creation of these new rights and the decision to deepen rather than replace the liberal approach to protection. To understand why these rights were created, I focus on the transformation of the policy agenda in the mid and late 1960s. The decision to give workers a right to safe and healthy work can be understood only in the context of the larger shift in the nature of American liberalism during the halcyon days of the Great Society. In this case, rank-and-file discontent over work combined with the radical visions of labor activists, environmentalists, and public-interest reformers to feed the issue of working conditions into the swelling tide of demands for reform.

The decision to retain the older forms of intervention was shaped by similar forces. The same reformers who sought to extend public authority failed to grasp the nature of the problem they hoped to solve. Caught in the grip of economism, the unions preferred to transfer responsibility for work safety to the state rather than challenge the power of employers at the workplace. The environmental and public-interest movements also failed to question the existing forms of public authority. Although these groups proposed a new view of the workplace—it was an "indoor environment"—and took rank-and-file participation more seriously than the unions did, they failed to grasp how capitalist relations of production structured work and shaped the problem of occupational health.

As a result, these groups demanded federal regulation, not a fundamental reorganization of work. They succeeded in placing the first issue on the agenda, but the second was not raised. Consequently, most of the debate over the OSH Act involved the details of program design rather than the relationship between capitalism and occupational safety and health. On this terrain, the legislation developed incrementally, building on previous efforts. Since these efforts were based on factory legislation, this approach provided the foundation for the new law.

The Transformation of the Agenda

Occupational safety and health became a political issue in the 1960s as a result of the intersection of a complex set of social forces, including rank-and-file discontent over work, union efforts to reform existing state programs, middle-class movements for environmentalism and consumer product safety, and White House interest in the development of a new policy agenda. Beginning as distinct movements in the mid 1960s, these social forces coalesced after 1966 to produce a groundswell of support for social regulation.

The "Labor Revolt" and Working Conditions

The 1960s were years of social protest, and although it is common to view the decade as one of black and student rebellion, industrial workers fought battles of their own, including a struggle against oppressive working conditions. This discontent did not lead to a movement by rank-and-file workers or the unions for the reform of the workplace. But it played a pivotal role in sparking and legitimating the efforts of a small group of Washington-centered health and safety activists, who translated that discontent into a movement for federal regulation of occupational safety and health.

Beginning in 1964, American workers became more militant and labor relations became more conflictual than had been the norm since the late 1940s. Increased rank-and-file discontent shows up clearly in the data on strikes and worker attitudes. In 1966 and the first half of 1967, the incidence of strikes reached a 10-year high. Wildcat or "during term" strikes, which are taken by many labor observers to be a proxy for discontent with working conditions, rose dramatically between 1967 and 1969.[1] Opinion polls suggest that the conditions of work played an important role in these developments: the University of Michigan Survey of Working Conditions reported that health and safety hazards headed the list of problems considered "sizable" or "great" by production workers in 1969.[2]

Nonetheless, the ideological and institutional arrangement of the accord resisted collective action around occupational safety and health. Workers were expected to ignore the workplace, and unions were encouraged to focus on economic issues rather than working conditions. Conforming to these expectations, the vast majority of worker protests over working conditions were spontaneous and disorganized. Wildcat strikes did not result in permanent organizations or even significantly different union strategies at the bargaining table.

Two genuine grass-roots health and safety movements did emerge from this ferment. Coal miners protesting health and safety conditions in West Virginia and Kentucky organized at the grass-roots level in 1969 and won a variety of concessions from employers and government. Working through the Black Lung Association and in conjunction with medical doctors and labor activists—initially against the wishes of the United Mine Workers—43,000 miners shut down West Virginia's

mines for three weeks. They also marched on the state capital to demand workers' compensation for black lung disease.[3] Ultimately, the
threat of a wider coal miners' movement forced congressional action,
helped to pass the Coal Mine Safety and Health Act of 1969, and
generated support for the OSH Act. Rank-and-file uranium miners in
Utah also attempted to defend themselves against workplace hazards. Frustrated with the delay and indifference of state workers'
compensation boards to their claims for disability payments for
radiation-induced lung cancer, they sought judicial relief and won
national media attention.

Organized Labor and Workplace Reform

For the most part, union officials remained indifferent to these demands. The AFL–CIO was not opposed to workplace reform; it simply failed to take it seriously. Neither George Meany, AFL–CIO president, nor Andrew Biemiller, Meany's chief aide and the federation's
head lobbyist, took social regulation in general seriously; the AFL–
CIO did not make occupational safety and health its first priority until
1970, when a bill was certain to pass.

Union leaders who were not skeptical or indifferent seem to have
been unaware of the hazards that workers faced. For example, despite the mounting evidence of radiation hazards and the union's
organizational interest in the issue, OCAW officials accepted the
Atomic Energy Commission's (AEC) position that there were no cumulative effects from low-level radiation exposure. The Steelworkers
(USWA), presented with evidence by a concerned researcher that
coke-oven emissions caused cancer, dutifully accepted the report
and then forgot it.[4] Some union leaders later admitted their role in
burying these issues. Testifying before the 1969 House Select Subcommittee on Labor, Steelworkers President I. W. Abel asked Congress to "make up for lost time . . . that we, too, have lost."[5]

Given the attitudes of the union leadership, the responsibility for
turning rank-and-file discontent into a union demand fell to a small
group of Washington-centered labor activists, public health reformers,
and interested bureaucrats in BOSH and LSB. Working on the margins of the labor movement, they helped make occupational safety

and health a national issue. Of these, the labor activists were by far the most important single influence. They were full-time unionists and lobbyists with access to union officials and the middle levels of the executive bureaucracy. Working with public health professionals in and around state agencies, they built a movement that folded rank-and-file discontent over work into the general movement for health and safety.

Three men played critical roles in building this movement. George Taylor, a staff economist in the AFL–CIO research department, came to the issue in the early 1960s. A few years later, Anthony Mazzochi of OCAW joined him. At the time, Mazzochi was an OCAW organizer; later he became its citizenship–legislative director, and eventually a union vice-president. In 1967 both men were joined by John J. Sheehan, an aide to I. W. Abel and a legislative lobbyist for the Steelworkers.

Although their efforts were not formally coordinated, a natural division of labor quickly developed among them. As the AFL–CIO's representative on health-related commissions and task forces, Taylor worked with mid-level bureaucrats in government and the labor movement. Mazzochi publicized the issue among the rank-and-file and argued labor's case to the wider public. Under his leadership, OCAW organized a conference on "Hazards in the Industrial Environment" and made links with environmental activists. Sheehan lobbied on Capitol Hill and served as organized labor's representative on top-level commissions and task forces.

Taylor proved adept at using the state as a platform for mobilizing support for worker health and safety. His first success occurred in 1965 when he helped to organize the task force that issued "Protecting Eighty Million American Workers," or the "Frye Report." This effort began modestly. When the PHS proposed to abolish the Division of Occupational Health, a small research agency, Taylor and a group of public health doctors in HEW sought to save it. In response to their activities, HEW appointed a task force to outline a new rationale for the beleaguered agency.[6]

The Frye Report proposed to turn the small agency into an empire and, in doing so, argued for a major change in the nation's approach to occupational safety and health. The PHS was urged to give the Division of Occupational Health major regulatory powers and a budget of $50 million a year. Most important, the report urged the PHS to take an aggressive approach to the problem and adopt a holistic view of worker health.

In its general statement of purpose, the report reminded the PHS of the New Deal vision of state-led social reform. "In America today," it stated, "it is not acceptable that any worker should pay with his health or his life for the privilege of having a job." It defined health hazards broadly to include damage from toxic chemicals and psychological stress. Moreover, it treated the workplace as an entry point for monitoring and controlling the general health problems of a majority of the population. In fact, it obliterated the line between occupational and other diseases. The national goal of the program was to "eliminate or control any factor in the work environment which is deleterious to the health of workers," including the "promotion of good health and well being." The workplace was simply one among many institutions that could serve as part of a national program of expanded health care.[7]

The Surgeon General rejected the report's ambitious recommendations, but Taylor and his colleagues had succeeded in bringing the issue to the attention of the AFL-CIO. The report also helped make public health professionals aware of the problem. From 1965 on, every discussion of occupational health in Congress and the executive branch had to respond, in some fashion, to its radical vision.

Taylor used the government's authority over atomic energy in a similar way: to focus union and public attention on the problem of worker health. There had been a major expansion of uranium mining in the United States in the late 1940s as part of the nuclear weapons program. In mining uranium, workers had been exposed to radioactive gases and dusts. As late as 1966, preventive measures to protect uranium miners were absent or inadequate. Despite success elsewhere with controlling airborne radiation with ventilation equipment, American firms resisted changing methods on economic, technical, and scientific grounds.[8]

The government suspected that exposure to radiation was dangerous. The PHS, the AEC, and state health departments in the Colorado Plateau (the lion's share of uranium mining occurred there) began studies of the problem in the 1940s. By 1960, the evidence strongly suggested a causal link between exposure to radon gas and lung cancer and the likelihood that many of the miners who had received large cumulative doses of radiation would die from it.[9]

None of the interested agencies acted to protect the miners. Regulatory jurisdiction over uranium mining was divided among four organizations: the AEC, the Bureau of Mines in the Department of the Interior, the PHS, and the labor department. The Bureau of Mines and the Department of the Interior and the DOL deferred to the AEC on national security grounds. The AEC deferred to its contractors. In effect, the subcontractors were left to regulate themselves. As a result, nothing substantial was done to alleviate the hazard.[10]

The federation and several individual unions protested the situation, but these efforts were halfhearted. Leo Goodman, a staff member in the AFL-CIO's social security division and an adviser to the UAW on atomic energy, attempted to generate support for regulation in the 1950s and early 1960s but was rebuffed by the union leadership and government. Taylor learned of the hazard as the AFL-CIO's representative to a task force on atomic energy; he also appealed to the federation to lobby the labor department to use its Walsh-Healey jurisdiction to protect uranium miners. Again, the AFL-CIO's response was perfunctory. When the Department of Labor resisted—national defense was involved, and the AEC tenaciously defended its jurisdiction—the federation backed off.

Eventually, the miners' own efforts, pressure from Washington-based activists, and national media attention forced the labor department to act. Assistant Secretary for Labor Standards Esther Peterson, a longtime ally of the labor movement, took a well-publicized visit to a western uranium mine in 1967. On her return, she urged Secretary of Labor Willard Wirtz to act. The DOL then exercised its jurisdiction despite opposition from the AEC. To be sure, the department's efforts remained limited. It deferred to the policies of the affected states on workers' compensation and accepted the AEC's recommendations on exposure levels. But the DOL did issue new exposure standards. Most important, the publicity surrounding the case generated interest in occupational health, and the reformers won a second victory in their effort to force government to take responsibility for the conditions of work.[11]

As they sought to draw the federal government into workplace safety and health, labor activists also lobbied their own unions to take

up the issue. Given the labor movement's general indifference to occupational safety and health, this step was critical to the movement's
success. Given the unions' indifference, it was also as difficult as the
effort to win federal support. Predictably, the first breakthroughs came
with the activists' own unions: OCAW and USWA. Then, using them as
platforms, and the political power of the Steelworkers as leverage,
they committed the AFL-CIO to reform.

Despite OCAW's reluctance to demand strict regulation of radon
gas, it did have a vested interest in health and safety because many
of the workers that it sought to organize were in dangerous industries.
That this could lead to union gains became clear when union organizers used the health-hazard issue to defeat an effort to decertify the
OCAW local at the United Nuclear Fuels plant in New Haven. After
that, Mazzochi was able to use OCAW as a platform for his wider
activities.[12]

Because of the ambitions of its president, the USWA also proved
receptive to making health and safety reform a political issue. Abel's
own fortunes were bound up with his claim to speak for the rank and
file, and he tried to use health and safety to make that claim compelling. The labor revolt challenged labor leaders such as Abel at the
same time that it threatened employers. National and local opposition slates appeared more frequently and were often successful in
defeating incumbents during the 1960s. The sitting presidents of the
American Federation of State, County, and Municipal Employees and
the International Union of Electrical Workers were upset by internal
revolts. Abel himself had defeated his predecessor in 1965 by appealing to rank-and-file discontent with the union bureaucracy.[13] But Abel
did not redeem his promises. He did not democratize the union or win
major economic concessions from industry after coming to office. In
1968 he faced a disillusioned rank and file and the threat that the
same kind of discontent that had brought him to power could unseat
him. Emil Narick, a relatively unknown staff lawyer, challenged Abel
for the union presidency and won more than 40% of the vote.[14]

Apart from internal reform or aggressive wage bargaining, Abel
had few ways to rebuild support among union members. Operating
under these constraints, he took the offensive and attempted to shift
the union's agenda. Like the politicians around him, who rushed to
endorse the new "quality-of-life" issues, Abel lectured the rank and

file about the importance of nonwage demands: environmental and occupational health and safety became union priorities. In 1968 the USWA became actively involved in air-pollution-control efforts in Pittsburgh, where its headquarters were located. It held a conference on air-quality control and urged the rank and file to become involved in the environmental movement. Abel also pressed worker health and safety in collective bargaining, and he took a personal and well-publicized role in lobbying for the OSH Act. Busloads of workers were brought to Washington to lobby their representatives and senators. To make sure that the entire membership got the message, the union showered the rank and file with reports of its efforts in their behalf.[15]

The federation was harder to convince. Urged on by Taylor, the AFL–CIO endorsed the Frye Report when it was issued. It petitioned the labor department to establish federal standards for uranium miners. It urged reform of the workers' compensation system and grants-in-aid to the states. The annual convention at Bal Harbour, Florida, in 1967 resulted in a formal resolution calling on the Department of HEW and the DOL to cooperate in a national program focused on occupational health. But George Meany and the Executive Council were more concerned about trade policy, repeal of Sec. 14b of the Taft–Hartley Act, and President Johnson's endorsement of a voluntary wage-restraint program. Occupational hazards remained a marginal issue, and when the PHS rejected the Frye Report, the federation acquiesced. It also accepted the AEC's position on the dangers of uranium mining.[16] In general, the issue of health and safety was left to those unions most interested in it.

The Steelworkers' interest in workplace regulation provided the stimulus needed to commit the federation to the movement for reform. After Walter Reuther withdrew the UAW from the AFL–CIO, the USWA became the largest and most important union in the federation. Most significantly, the UAW's departure gave Abel control of the federation's Industrial Union Department (IUD), the organizational basis for coordinating political activity among unions in the mass-production industries. In keeping with his strategy to make occupational safety and health a union priority, Abel used this position to commit the federation's resources to lobbying for federal regulation of working conditions. In 1968 Sheehan became the lead lobbyist for an occupational safety and health bill, and Abel mobilized the federation's political resources behind the movement for reform.

Middle-Class Movements

As they did in the movements for consumer product safety and environmental regulation, middle-class reformers and radicals played a role in the demand for workplace regulation. Their role is easily exaggerated, however, and it is important to consider carefully the part they played. Occupational safety and health reform never became a principal demand of the public-interest or environmental movements. The issue never caught the public's attention or attracted the media coverage that air and water pollution did.[17]

The middle class fed the demand for reform in other ways. By calling attention to noneconomic issues and the indifference of many industrial corporations to the environmental effects of their market activities, writers such as Rachel Carson (author of *Silent Spring*) and reformers such as Ralph Nader prepared public opinion for workplace reform. At the same time, professionals and bureaucrats in state agencies joined labor activists to legitimate the idea of federal regulation of work. Finally, in coalition with labor groups, environmental and public-interest lobbyists intervened at key junctures in the legislative battle over the OSH Act and helped mobilize wavering senators and representatives.

The public health professionals and medical doctors who worked with unions and rank-and-file workers were particularly influential because they were able to counter the antistatist views of the industry-oriented private professional organizations. Some doctors worked with rank-and-file movements to gather basic data on health hazards and do the kinds of research that corporations, state agencies, and private professional groups did not do. Dr. Selikoff's study made asbestos a national issue. Drs. Hawey Mills and Lorin Kerr did the same for black lung. Dr. I. E. Buff helped publicize the hazards of cotton dust. Other doctors worked within government to press agencies to take worker health more seriously. Their role was necessarily more circumspect, and they were often more moderate in their political orientations than the doctors who worked with rank-and-file activists. Nonetheless, they played strategic roles, as the effort to save the Division of Industrial Hygiene illustrates. Other public health doctors and environmental activists educated unionists about the relationship between the work environment and occupational diseases. Given the labor movement's traditional emphasis on safety rather than health, these efforts were indispensable.[18]

Public-interest and environmental organizations and activists also played a key role in the congressional debate over OSHA. Their presence suggested that workplace reform was in the public interest rather than in the "special" interest of the unions. The labor activists knew they needed to make this case and worked to build bridges to these other movements. Aside from organized labor's long-standing allies, legislators were notoriously unsympathetic to "union" issues that seemed to serve the organizational needs of the labor movement. They could be swayed, however, if convinced that more general "worker" interests were at stake.

These efforts to build alliances paid off. In 1968 a coalition of more than a hundred labor, consumer, religious, and environmental groups formed to lobby for the OSH Act, giving the impression of widespread support for this reform. At the climax of the legislative battle, the environmental movement publicly urged Congress to adopt the strongest possible bill. As hoped, its support helped the health and safety reformers rebut industry claims that the Democratic bill served the special interests of unions rather than the general interests of workers. Environmental Action issued a letter, signed by a group of notable scientists, that explicitly drew occupational safety and health under the broader umbrella of the new environmental consciousness. "The in-plant environment," it stated, was "merely a concentrated microcosm of the outside environment. The environmental health hazards that workers face affect the entire population."[19]

Ralph Nader included the movement for workplace reform under his public-interest banner. In the summer of 1969 he assigned the second generation of raiders to investigate how the Departments of Labor and HEW handled occupational health and safety. The raiders uncovered evidence of lax enforcement and cozy relationships between state and industry, and their reports were widely circulated. As the environmentalists had done, Nader's efforts served to cast the issue in the widest possible terms and force industry onto the defensive.[20]

The White House Interest in Social Regulation

From 1966 on, President Lyndon Johnson and his closest aides also worked to make occupational safety and health a political issue. The Johnson White House was interested in a post-civil rights policy agenda and, in particular, new symbols of reform in an increasingly cost-conscious political environment. It seized on social regulation and, within this broader frame, occupational safety and health.

It is hard to overestimate the importance of political entrepreneurship of this sort to the success of social regulation and workplace reform. The postwar accord had little to say about health, safety, and the environment. As was true with the workplace safety and health activists, the consumer and environmental movements were small and poorly organized. Widespread public support for air- and water-pollution control did not emerge until the second half of the 1960s. Militant and politically sophisticated environmental groups—for example, the Friends of the Earth, the Environmental Defense Fund, and the Natural Resources Defense Fund—were founded in the late 1960s and early 1970s. The consumer lobby amounted to one organization —the National Consumers League—with one full-time staff person until Ralph Nader organized Public Citizen in the late 1960s. By then, however, the executive branch had supported and Congress had passed legislation covering air and water pollution and motor vehicle and product safety.[21]

The White House's interest in social regulation was natural and logical in the political environment of the mid and late 1960s. First, the party's majority status increasingly rested on new middle-class voters. Educated professional, managerial, and technical workers who had previously voted Republican were shifting toward the Democratic party, and the party could benefit from signs that it was responsive to their concerns.[22]

Social regulation was also a relatively costless way for the Democratic party to demonstrate its continuing commitment to reform. After

1966, growing concern over the fiscal demands of the Great Society made Congress increasingly cost conscious. "Big ticket" programs were scrutinized more carefully. Health, safety, and environmental regulation, however, did little to increase the federal budget. The administrative costs of running a regulatory program were relatively low; the important costs were borne by groups in society. This also appealed to the Johnson administration.[23] Finally, Congress had taken the lead in this area and had all but forced Johnson to respond if he was to take any credit for consumer and environmental legislation.

The administration's interest in occupational safety and health regulation makes sense given the Democratic party's relationship to organized labor. Although the Democrats courted new middle-class voters, the party's political fortunes continued to rest on the Democratic affiliation of blue-collar voters and the organizational commitment of the union movement. The 1966 midterm elections suggested that white workers had begun to rethink their traditional affiliations. A poll of union members commissioned (and then suppressed) by the AFL–CIO confirmed this impression: resentment of welfare state policies directed at the poor, fear of urban unrest, and opposition to tax increases to pay for new social programs fed working-class discontent with the party.[24]

The Democrats' relationship to organized labor was also problematic. The party was indebted to the unions. President Truman had promised to repeal the Taft–Hartley Act in 1948; Adlai Stevenson had repeated the commitment in 1952; and Lyndon Johnson promised to lead a movement to repeal Sec. 14b in 1964. The unions did their part and helped to turn the Great Society into law in 1965. They expected to be paid back with labor-law reform and were encouraged by large Democratic majorities in both houses of Congress. But Congress rebuffed the unions in the mid 1960s. Although organized labor remained on good terms with President Johnson, his failure to secure the repeal of 14b strained the relationship. Some in the labor movement complained that Johnson had traded labor-law reform for southern Democratic support for civil rights legislation. Other Johnson positions, including his endorsement of legislation restricting picketing on construction sites and his campaign for voluntary wage restraints, put added stress on his alliance with the unions.

In this context, occupational safety and health was an opportune response to all these forces. It simultaneously affirmed the party's commitment to the new politics of middle-class reform and the traditional concerns of organized labor, concerns that were often seen as mutually exclusive if not directly competitive. As one kind of social regulation, worker health and safety was an important symbol of the administration's commitment to the public-interest and nascent environmental movements. Unlike more traditional forms of labor legislation, it spoke to the interests of all workers, whether unionized or not. Carefully packaged, it could also be used to pacify organized labor.

There has been a good deal of speculation about how occupational safety and health came to Johnson's attention. The standard explanation is that a presidential speechwriter who had a brother in BOSH managed to slip a few references to the issue into Johnson's speeches. This somehow caught the president's eye. Although accurate in one respect—the fraternal connection was real—this account is misleading. It suggests that the policymaking process in this instance was more serendipitous than it actually was, and it argues for the view that the issue and the OSH Act were poorly understood or thought out.[25] Actually, occupational safety and health entered the White House policy agenda through another route. That route was indirect, but the issue was taken much more seriously than previous accounts have indicated.[26]

Occupational safety and health became part of the administration's policy agenda in an effort to develop new "quality-of-life" issues to take to the electorate when Johnson ran for reelection in 1968. Workplace reform was not the central issue, to be sure; consumer product safety was more important. Johnson proposed the Highway Safety Act in his 1966 State of the Union Message and endorsed four other pieces of consumer legislation that year. Air- and water-pollution control were also higher priorities.

Nonetheless, Johnson's aides searched widely for related issues, and occupational safety and health came up in that context. In 1966 Joseph Califano, Johnson's chief domestic policy adviser, organized a search for new quality-of-life issues under the general rubric of "accident prevention." All interested departments were asked to identify areas of responsibility that fit this frame. The Department of Labor mentioned "occupational safety" along with a host of other topics. While the DOL did not press it, Califano thought it a natural part of the new emphasis. In this way, workplace reform became part of the White House's policy agenda.[27]

Given that organized labor was more interested in other issues, Johnson had to work hard to sell workplace regulation to the unions. In his first public reference to the issue, in May 1966, he urged the labor movement to rethink its agenda. Speaking to a group of labor journalists, he asked the unions to give up their preoccupation with "bread and butter" issues and "join with us in the effort to improve the total environment,"[28] and his administration went to work on a bill. For the next year and a half, the Bureau of the Budget (BOB), under Califano's direction, pressed both the DOL and HEW to propose a strong occupational safety and health program to be included in the president's 1968 program for labor.

In the end, Johnson's efforts proved decisive in placing the issue of workplace safety and health on the policy agenda and legitimating the demand for federal intervention. While labor activists were still trying to convince the federation to make OSHA a priority, the administration introduced its bill to extend federal regulation to most of the labor force. Moreover, it defined the issue in exactly the way that labor activists had pressed organized labor to do, giving it the broadest possible appeal.

Speaking for the administration, Secretary of Labor Wirtz placed the bill in the quality-of-life framework that Johnson had urged on the unions in 1966. It was, he said, a victory for a new view of politics, a view in which social values took precedence over economic values. There was, Wirtz told Congress, a new tendency to pay attention to "human concerns"—to "measure progress in qualitative as well as quantitative terms"; to forsake "human sacrifice for the development of progress." Our priorities, he maintained, had been permanently reshaped by economic advances: "Now higher value is placed on a life, or a limb, or an eye." Wirtz was explicit about the underlying choice in the proposal to extend federal regulation in this way. The administration's bill, he claimed, "asserts the absolute priority of individual over institutional interests and of human over economic values. . . . " In this fashion, it "faces up to the most basic issues of contemporary thought and values."[29]

Why Business Failed to Stop OSHA

Most affected business interests rejected the idea of federal regulation of work. Yet the Johnson administration introduced its bill in 1968, and Nixon recommended a revised version in 1969. In short, the policy agenda had been dramatically altered against the opposition of a broad spectrum of business lobbies. Almost all theories of business power in America suggest that this reform should have failed, particularly against the concerted opposition of so many business groups. Why, then, did it succeed?

Strategic errors by the business lobby undermined industry's ability to control the agenda in the 1960s. Most important, business lobbyists refused to admit the need for change and chose to deny the problem and defend the existing private professional groups and state and local regulatory efforts despite their obvious failures. They blamed the workers themselves or questioned the motives of the labor movement. The influx of younger workers and the breakdown in labor discipline caused increased accident rates, they maintained. The unions were trying to increase their power at the workplace, not to protect workers from hazards. The Chamber of Commerce spoke for this coalition, and its position suggests the nature of its opposition. The unions, the chamber suggested, were trying to create a labor "czar" with life-and-death powers over industry. *Nation's Business*, the chamber's magazine, warned its readers that formerly unemployed welfare clients would return as OSHA inspectors to seek revenge on American capitalism.[30]

Several more positive approaches were available to employers. They might have offered to implement a federal program through joint health and safety committees organized in such a way that they could be dominated by employers. They could have acknowledged the toll taken by occupational hazards and argued for liberal compensation to injured workers funded by public moneys, or for in-plant occupational health clinics subsidized by the state. They might have proposed a regulatory system that exempted "safe" firms and concentrated on the most hazardous industries and companies.

There was also general support among many corporate leaders for social reform in this period. Many executives understood that the market had failed to protect society against the hazards of industrialism and believed that reform was inevitable. Henry Ford II of Ford Motor Company, for example, urged his colleagues to accept that "the terms of the contract between industry and society are changing." He told them to lobby instead for quality-of-life programs that were compatible with capital investment and that relied on market incentives to translate social goals into public policy.[31] Sol Linowitz of Xerox Corporation advised corporate executives to lead the quest for social justice by internalizing the new values.[32]

But employers reacted defensively to demands for workplace regulation. Even many of the most "safety conscious" firms fell into this trap. They tried to strike a more conciliatory tone, but their basic position was similar to that of the Chamber of Commerce. The American Iron and Steel Institute (AISI), speaking for the steel industry, argued that "basic progress in occupational safety and health has been made, primarily, on the basis of voluntary action." There was no place for "compulsion"in a comprehensive safety and health program. "The really important progress in occupational safety and health," their lead lobbyist suggested, "would require far more consideration of the man rather than the environment."[33] The chemical industry took great pride that "continuing strides in occupational safety and health have been achieved through the voluntary efforts of businessmen, both individually and through trade associations."[34] The former president of the Industrial Medical Association summarized the views of professionals working for the larger firms: "In many of the major industries the programs in occupational safety and health are successful, are well advanced, and have been developed to the point where the most important remaining problem is human failure."[35]

Two factors explain this strategic failure. First, employers traditionally saw occupational safety and health as a potentially explosive labor–management issue. As the preceding chapter indicates, firms sought to control conflict over working conditions in order to maintain their control over the labor process. From this vantage point, employers considered organized labor's effort to use federal power to regulate the workplace as a tactic in the ongoing power struggle between employers and employees. Unions, they believed, sought occupational safety and health regulation as another resource that might be brought to bear in collective bargaining. The threat of a federal inspection by an agency sympathetic to organized labor might cause employers to make concessions on other issues. As the AISI lobbyist confided to Congress, "practical operating managers know" that "safety or health issues are frequently alleged merely to build up a case" against employers.[36]

This strategic failure was also a result of an organizational lacuna that had developed within the business community in the 1960s. No peak association existed that could speak for employers as a whole and negotiate these issues with organized labor and government. This was particularly important because employers were divided among themselves. Most important, different firms faced different cost and control problems: some industries were more hazardous then others; some firms upgraded their control technologies as they invested in new plant and equipment; others did not reinvest in new machinery and, consequently, worked with less-safe facilities.

These differences might have been resolved by class-conscious leadership, but none emerged. The insurance industry was best situated to attempt it. Companies that underwrote workers' compensation had an economic interest in safe workplaces. They also had reliable information about the hazards of work and the problems that reformers faced in seeking to change employers' behavior. But the insurers chose to defend their own interests rather than those of firms as a whole. Instead of taking a leading role in the debate over regulation, the insurance industry focused its efforts on defending the states' workers' compensation programs against proposals to reform or possibly federalize them.

The two preeminent liberal business organizations—the Committee for Economic Development and the Business Council—failed to take any position at all on occupational safety and health. They concentrated on economic policy in the mid 1960s and were generally unprepared for health and safety legislation. The private health and safety groups functioned as policy think tanks and public relations organizations for the safety-conscious firms, but they also failed to respond flexibly to the challenge. The NSC staff recommended that the organization endorse a strong federal program but the executive board overruled them and lobbied for a limited program that left enforcement to the states.[37] Other private health and safety groups opposed federal standard setting altogether; the AIHA, AMA, and

ASSE rejected the concept of mandatory standards. As a result, the Chamber of Commerce took over the leadership of the movement against reform and designed a defensive strategy poorly suited to the political climate of this reformist period.

Designing the Program

As is often true in American politics, once the agenda was transformed, the conflict over workplace reform focused on the details of state intervention rather than basic principles. At this level, three related issues were paramount: (1) the precise relationship between federal and state power, including the fate of the workers' compensation system and the public health program proposed by the Frye Report; (2) the role that private professional and standard-setting groups would play in the program; and (3) whether standard setting and enforcement would be concentrated in a single executive department or divided among a number of agencies.

The Decision for a Federal Program

Although the business lobby, the private professional groups, and the state agencies fought it, the decision to federalize the program was made easily. The trend everywhere in government at this time was toward national power; the Great Society was premised on it. As Chapter 2 detailed, there were no compelling reasons to change direction in this instance; it was clear from the evidence that most states had done a terrible job of regulating occupational hazards. It was also clear that few states could be counted on to improve their own programs on their own. Indeed, the states had resisted federal grants-in-aid designed to upgrade their activities. As a result, the Johnson administration assumed from the start that the new program would expand federal authority.

Nonetheless, it took a year of pushing and hauling to turn this decision into a concrete policy proposal. Not surprisingly, HEW's and DOL's institutional lethargy was hard to overcome. The secretaries of

both departments were not attentive to the problem of working conditions, and neither responded enthusiastically to Califano's requests.

Taking its lead from the Frye Report, the White House first gave
HEW responsibility to develop a program, but the department was
indifferent.[38] Despite the Frye Report's recommendation that HEW
regulate the workplace as part of an expanded public health effort,
neither HEW nor the PHS was interested. Of course, Secretary of HEW
Wilbur Cohen and, later, Secretary John Gardner had other priorities;
their department was responsible for developing and implementing
many of the Great Society's new social welfare programs, from education to social security. As a result, they had more than they could
easily handle and were willing to leave workplace safety and health
to the labor department.

Given its long-standing jurisdiction in the area of worker health, the
PHS might have been expected to press for a piece of the new program; BOSH was authorized to conduct studies, undertake field investigations, and mount demonstration projects to help detect occupational disease. It also advised the private standard-setting bodies.
The PHS did not rise to the challenge. It generally eschewed statist
solutions to public health problems and supported states' rights in
health-related areas. Even when concern over the health effects of
smoking and air pollution catapulted the agency into the public
limelight, it recommended against legislation that preempted local
jurisdiction. This pattern held. Since the Frye Report contradicted the
PHS's traditional approach, the Surgeon General rejected its recommendations.

Jurisdictional disputes between the Departments of HEW and Labor
further hampered efforts to develop a bill. Although the PHS rejected
the idea of federal regulation, it jealously guarded its organizational
prerogatives and resisted efforts to augment LSB's role. Despite orders
from the White House to come up with a "strong and imaginative"'
program, several interagency task forces established in 1966 fell victim to disputes between HEW and Labor and HEW's general indifference. Dr. Phillip Lee of HEW, the head of one of these efforts, rejected
the Bureau of the Budget's request for a bill. "The knowledge and
resources," he observed, "are simply not at hand, either at the Federal,
state or local level."[39]

Frustrated by a year of delays, the White House broke this impasse by taking the program away from the Departments of Labor and HEW and giving it to the BOB. Through the fall of 1967, BOB officials worked on the program quietly. Top officials at HEW and the DOL were shut out from the deliberations, and middle-level staffers were called in to provide BOB with the details of existing programs. The bureau resolved the jurisdictional issues itself. For a variety of reasons, including organized labor's preference for this arrangement, the program was given to the DOL. Budget decided to extend the program to as many private-sector workers as constitutionally possible, to give the DOL standard-setting powers, to secure compliance through penalty-based inspections, and to reject the Frye Report's call for publicly funded in-plant clinics. Johnson's advisers were apprehensive about civil penalties for recalcitrant employers, but they agreed to all of BOB's recommendations. Once these decisions were made, Secretary of Labor Wirtz and Assistant Secretary Peterson were called to the White House and asked to propose legislation to implement the program that the bureau had designed.[40]

The administration also decided against reforming the workers' compensation system, although that program's manifest failures made reform logical. Organized labor considered this as important, if not more important, than a federal regulatory program, and Johnson's own advisers were attracted to the idea of rehabilitating the system to provide economic incentives to employers to improve working conditions. But Wirtz met with insurance industry representatives to test the waters and returned to recommend two very limited reforms: a grants-in-aid program to improve state research and administration, and a congressionally appointed National Commission on Workmen's Compensation.[41]

Johnson's advisers clearly understood the economic and political interests at stake and the problems they faced if they chose to tackle workers' compensation directly. After reviewing the history of the program, both the Council of Economic Advisers (CEA) and BOB concluded that comprehensive changes were unlikely to succeed in Congress. Gardner Ackley, chair of the CEA, wrote Califano that "even innocuous government efforts to improve the system have been vigorously assailed and strongly resisted as precursors to a Federal 'take-over' of the system." "Given the entrenched power of the defenders of the status quo," he concluded, minor gains were all that could be expected. Budget concurred. "In our view," the agency argued, "the basic problem is the unwillingness of State legislatures and employers to pay the large costs involved in establishing adequate benefits." Although unenthusiastic about it, BOB endorsed the idea of the commission in the hopes that a public study might stimulate future reform.[42]

After more than a year of internal debates, Johnson finally proposed the OSH Act in his Manpower Message in January 1968. Introduced into Congress as the O'Hara-Yarborough bill, it followed the outlines of the BOB plan and gave the Department of Labor the authority to set and enforce federal health and safety standards for almost all private-sector employees. In many respects, this version of the bill was rudimentary: it lacked provisions for worker rights; the standard-setting process was poorly specified; the federal-state relationship was left ambiguous. But O'Hara-Yarborough foreshadowed the final act. Standard-setting and enforcement powers were concentrated in the same agency. The secretary of labor was given broad discretionary power to adopt standards that departed from those developed by private organizations, and standards did not have to take costs into account.

Where Should the Agency Go?

From the employers' point of view, the situation did not improve dramatically with the election of Richard Nixon. Many business groups thought that, as a Republican and an advocate of what he called the New Federalism, Nixon might oppose federal regulation of work. And Nixon was much more sympathetic to employer interests: he helped business lobbyists prepare and propose a series of bills that would have substantially altered the Democratic program. But the Nixon administration never seriously considered abandoning the idea of a federal regulatory program.

To the Nixon administration, two circumstances argued against opposing federal regulation of work. To begin with, Nixon's political options were limited. Occupational safety and health was already on the agenda when Nixon took office. The first Democratic bill had died in Congress after Johnson announced that he would not run for re-election. But, as Nixon's transition advisers warned him, the issue remained alive, and many legislators supported reform. The coal miners' protests and congressional deliberations on the Coal Mine Health and Safety bill had sustained interest in occupational hazards. The Democrats intended to introduce a revised version of O'Hara-Yarborough in the new session of Congress; they also seemed to have enough votes to pass it. Thus the Republicans had to propose their own bill if they were to play a central part in crafting the new program.[43]

In addition, workplace reform appealed to the Nixon White House for some of the same reasons that it appealed to the Johnson administration. Like Johnson, Nixon courted blue-collar voters; racial conflict and white backlash made it possible to compete with the Democrats for their support. As Kevin Phillips's *The New Republican Majority* suggested, a political strategy that targeted this group could make the Republicans the majority party for the first time in 40 years. Nixon's advisers took Phillips's advice and developed just such a "blue-collar strategy." Nixon's ability to appeal to the material interests of workers was limited, however. His economic policies—he engineered a recession to reduce inflation as soon as he entered office—had alienated organized labor. Nonetheless, support for OSHA, carefully qualified, could contribute to the pursuit of working-class votes; it could symbolize the administration's concern for the "silent majority." Nixon made this point clear when he signed the OSH Act. It was, he claimed, "probably one of the most important pieces of legislation, from the standpoint of 55 million people who will be covered by it, ever passed by the Congress of the United States."[44]

Once it became clear that federal regulation was inevitable, the struggle over reform focused on the specific institutional arrangements to implement the program. Here, employers had several concerns. They did not want the Department of Labor to set standards; they did not want standard setting and enforcement concentrated in the same agency; and they wanted costs taken into account in setting exposure levels.

Actually, a small group of safety-conscious firms and private-sector health and safety professionals first raised the issue of where to locate the new program in 1967 during two sets of meetings with organized labor and the Johnson administration. Seeking to reach a labor-management consensus before the submission of a bill, they were willing to concede the issue of federal authority if the administration would agree to locate the program somewhere other than in the DOL. A Department of Labor Task Force on Occupational Safety provided the forum for the first set of meetings. The second set, organized by Dr. David Goldstein, medical director of the *New York Times* and president of the Industrial Medical Association, occurred behind the scenes and looked promising. A substantial number of safety-conscious companies took part, including Kodak, Du Pont, American Telephone & Telegraph, and General Electric, and all agreed on the principle of federal regulation. But both sets of meetings failed to resolve the issue. The business representatives refused to accept a program lodged in the Department of Labor, and organized labor and the Johnson administration refused to relocate it.[45]

The Nixon administration's effort to secure a bill that protected employer interests also focused on where to locate OSHA. The Department of Labor, charged with developing the Republican bill, found an ingenious solution. An independent board of professionals would set standards, the DOL would inspect workplaces, and the courts or some other agency would determine fines and hear employer appeals.

Given the administration's unwillingness to challenge the principle of federal authority, the Nixon plan made sense from a business perspective. Aggressive implementation would be hampered by the division of authority among a number of agencies. Since the vast majority of health and safety professionals and almost all of the occupational health and safety organizations were sympathetic to the employer's point of view, standard setting was likely to compromise employer and employee interests. The courts, in turn, were more likely to be sympathetic to employers' property rights than the DOL would be.

Most business groups and private professional organizations recognized the logic of the approach and fell in line, but some business and professional groups argued for further limits on federal authority. The Chamber of Commerce asked that a joint federal–state program be set up, to be directed by a National Advisory Board drawn from private professional organizations. The NSC suggested that, whenever possible, standards be based on existing consensus standards and that future standards development be left to private organizations. It also urged that an advisory board be created, drawn from "the involved industries and groups," to oversee the standard-setting process. Like the Chamber of Commerce, it wanted enforcement left to the states.[46] In a March 1969 memo to the DOL, Secretary of Commerce Maurice Stans summarized the remaining reservations. Employers were concerned, he wrote, that the administration adopt an approach that restricted federal standards to privately developed proposals and subjected standards to mandatory economic review procedures to control costs.[47]

Secretary of Labor George Shultz and Undersecretary John Hodgson, in charge of drafting the administration bill, were sympathetic to these proposals but balked at the most restrictive ones. The DOL endorsed the idea of maximizing the states' involvement in enforcement and allowing the board to use private standards where they existed. But it rejected limits on the board's power to adopt and revise standards on its own, and rejected a statutory commitment to economic and technical feasibility.[48]

Eventually, the White House crafted three different bills designed to protect employer interests. The first, Javits–Ayres, created the tripartite division of authority referred to above. Standards were to be set by a National Occupational Safety and Health Board of experts; the Labor Department was to conduct inspections; the courts were to enforce DOL citations. This bill also limited federal authority in other ways. The board was required to promulgate ANSI and National Fire Protection Association standards—"national consensus standards"— where they existed. The Democrat's "general-duty" clause that required employers to provide healthy and safe workplaces was replaced with an obligation to conform only to board standards. Finally, employers were allowed to depart from board standards if they provided workers with conditions that were "substantially equal" to those mandated by board rules.

The second administration bill, the "Steiger substitute," was worked out in concert with AFL–CIO officials, Representative William Steiger (R-Wisc.) of the House Labor Committee, and several committee Democrats. Like Javits–Ayres, it lodged standard setting in an independent board. But it shifted enforcement from the courts to the board, increased worker rights to information, and provided for research on worker health.

These concessions were designed to win the support of moderate Democrats and the more cautious union officials and thereby split the forces supporting reform. They almost succeeded. The federation helped draft and then recommended the Steiger substitute. Unenthusiastic to begin with, Biemiller, Meany's assistant, counseled Sheehan of the USWA that a stronger bill would be defeated on the House floor, where southern Democrats and Republicans had more influence than they did on the labor committee. The subcommittee chair, Representative Dominick Daniels (D-N.J.) and the full committee chair, Representative Carl Perkins (D-Ky.), both long-time labor allies, agreed with Biemiller's position. Only Sheehan's opposition prevented the Democrats on the committee from recommending the Steiger substitute in place of the original Democratic bill.[49]

The third administration compromise located standard setting and enforcement in the board but substantially augmented health-hazard monitoring and worker rights to participate in enforcement. The negotiations that led up to this bill included Democrats and Republicans on the House Labor Committee and Nader aide Gary Sellars, and were personally supervised by Hodgson, who had replaced Shultz as secretary of labor. In this compromise, the administration sought to split the public interest–union coalition by appealing to the Naderites' interest in institutional arrangements that facilitated citizen participation in administrative regulation. Again, the administration almost succeeded. Nader praised the draft bill; he was personally willing, he reported, to move OSHA from the DOL in return for "some of the most ingenious self-enforcing provisions in any regulatory law."[50] Once again, however, the USWA rejected the compromise, and this bill also failed in committee.

Congress finally passed the union–Democratic bill in the fall of 1970; the business campaign against OSHA had failed. As Biemiller and Perkins had predicted, there were problems on the House floor. The conservative coalition of southern Democrats and Republicans rejected the Democratic bill and passed the Steiger substitute. But, as Sheehan had gambled, the Senate and the conference committee sided with the unions. While the OSH Act has Steiger's name on it, it contains almost none of his provisions. The Williams–Steiger Act is far stronger than the original O'Hara-Yarborough bill. Each new Republican proposal had ceded more ground than the last. In late 1970, faced with the choice of vetoing or accepting a bill that included almost all of organized labor's demands, Nixon signed and took credit for it. American business was forced to accept a program that it had unconditionally rejected in 1968.

Even the more carefully targeted efforts to win industry-specific concessions were largely unsuccessful. The petroleum industry, concerned about the costs of comprehensive recordkeeping and information gathering, failed to limit standards to situations where substances could be easily monitored. The construction industry failed to win exemption from the program. The steel industry, concerned that inspections would be used by workers to pressure employers in collective bargaining, was only partially successful in limiting worker rights to participate in enforcement.[51]

A Radical Liberal Reform

The Occupational Safety and Health Act of 1970 is a remarkable piece of legislation. From industry's point of view, it represented the worst of both worlds. White House efforts to win Democratic support for the board led House Republicans to accept several provisions dealing with worker rights—including the right to participate in inspections, have access to information about citations, and contest agency actions—that remained in the act even after the conference committee eliminated the board.

More important, the OSH Act codified a new, more radical, vision of worker rights. Conventional liberal ideology linked health and safety to individual, voluntary action in markets. But the OSH Act created a universal and substantive right to safety and health. Employers could not buy the opportunity to risk workers' health and safety, despite some workers' willingness to sell it. Moreover, this right was enjoyed by all workers regardless of their market position, income, or occupation. In sum, the act promised all workers a minimum level of health and safety regardless of the extent to which they were politically and economically organized, their income, or their market position.

Specifically, the act did the following:

■ Empowered the secretary of labor and the agency to which the secretary delegates responsibility (i.e., OSHA) to set and enforce standards governing the conditions of work of all employers except federal, state, and local government agencies.[52]

■ Created the National Institute for Occupational Safety and Health (NIOSH) in HEW to develop and recommend occupational safety and health standards, to compile and publish a list of toxic substances, to conduct research and experimental programs, to carry out hazard evaluations, and to promote the training of occupational safety and health professionals.

■ Obligated employers to comply with occupational safety and health standards issued by the secretary of labor. It also created a general employer duty "to furnish to each of his employees employment and a place of employment which are free from recognized hazards that are causing or are likely to cause death or serious physical harm to his employees" (Sec. 5a). Employers were also obligated to maintain records on worker injuries and illnesses.

■ Established civil and criminal penalties for violations of the act and the rules and regulations developed to implement it. These include fines of up to $1000 for most violations, up to $10,000 for willful or repeated violations, and up to $1000 per day for the failure to correct cited violations; fines of up to $10,000 and up to six months' imprisonment, or both, for willful violations that cause death to an employee; fines of up to $1000 or imprisonment of up to six months, or both, for giving unauthorized advance warning of inspections; and fines of up to $10,000 and imprisonment of up to six months, or both, for knowingly making or filing false statements in response to the information-reporting requirements of the act.

■ Created a number of employee rights to become involved in the administrative and enforcement activities of the act. These include rights to participate in standard setting, workplace inspections, and the monitoring of hazards; to have access to information about hazards and agency findings; to appeal certain agency rulings to a newly created Occupational Safety and Health Review Commission (OSHRC) and to the courts and to oppose certain kinds of employer appeals; and to be protected from employer discrimination for exercising these rights.

■ Required the secretary of labor, when issuing standards that dealt with toxic materials or "harmful physical agents," to "set the standard which most adequately assures, to the extent feasible, on the basis of the best available evidence, that no employee will suffer material impairment of health or functional capacity even if such employee has regular exposure to the hazard dealt with by such standard for the period of his working life" (Sec. 6b[5]).

This last provision was exceptional and has since become the subject of intense controversy in and out of the courts. Many conservatives, industry representatives, and some policy analysts insist that Congress could not have actually meant to provide protection regardless of cost and have tried to read some sort of cost–benefit test into the act's feasibility language. But the legislative history is clear on this point. The act contains only one reference to feasibility, and that is much vaguer than industry wanted. This was a major defeat for the steel and chemical industries in particular, which had lobbied Congress to require that OSHA standards be economically feasible. The unions objected on the grounds that such a provision would allow employers to appeal every standard and enforcement action, and the Democrats agreed. The industry proposal was rejected in committee.[53]

At the last moment, Senator Jacob Javits successfully introduced the reference to "feasibility" in the health standards section, and much has been made of this. But the bill's proponents and opponents knew that Javits's reference did not meet industry objections. According to Representative Perkins, chair of the House Labor Committee, the bill, as written, provided a "congressionally recognized right to every man and woman who works to perform that work in the safest and healthiest conditions that can be provided."[54] Republican critics agreed. Senator Peter Dominick (R-Colo.) was the author of several industry-oriented amendments that were defeated on the Senate floor. His comments, though hyperbolic, reflect industry and Republican anxiety about the act. "It could," he claimed, "be read to require the Secretary to ban all occupations in which there remains some risk of injury, impaired health, or life expectancy ... the present criteria could, if literally applied, close every business in this nation."[55]

Business did not lose every point. Congressional Republicans and southern Democrats were responsive to business lobbying, and the unions were forced to accept four last-minute compromises. They gave up what industry called the "strike with pay" clause that guaranteed compensation to workers who walked off hazardous jobs. The secretary of labor was denied the authority to shut down plants on his own authority. The act does not require the secretary of labor to hold a representation election to select an employee to accompany the inspector where no union exists. The Senate added OSHRC to provide employers with an independent forum in which to appeal citations.

In addition, state governments successfully lobbied for a joint program, although the states' role was more limited than they wanted. From the start, all bills allowed the states to retain jurisdiction over existing programs, or develop new ones, if those programs met federal standards. Business support for state-level regulation helped the states make their case. Employers hoped that a joint program would reduce their costs. As one steel industry lobbyist later put it, "Everyone knew that the State commissions were in bed with industry and everyone expected that the states would start up plans as soon as this passed." The state programs would, he expected, be a "safety valve" for employers.[56] Having insisted on federal standard setting, the Nixon administration also felt obliged to grant at least some of the demands of the states, and always responsive to local interests, Congress concurred. Subject to federal supervision, the act provided states with subsidies to develop and maintain their own programs.

The Limits of Reform

Despite the worker rights granted in the OSH Act, it is important to recognize the problems created by the decision to deepen rather than alter the existing approach to regulation. Workers had failed to challenge capitalist control of the labor process. Middle-class reformers had also failed to grasp the link between economic structure and occupational hazards. Indeed, Nader's account of the problem was remarkably conventional and sanguine about the relationship between capitalism and work. He told Congress that "one can easily envision a state of industrial activity where most of these injuries have been eliminated, and still have industry making just as much profit and if not more than they are now." The problem was simply that workers' interest in health and safety was "not being represented, represented adequately, by a special interest group." This was, he said, "the crux of the problem."[57]

Given these demands, the state had turned to conventional forms of factory legislation rather than require institutional changes in employer–employee relations. In fact, the act created a particularly narrow range of policy instruments. The state-level workers' compensation system was left in place. The public health orientation of the Frye Report was rejected. Employers were not required to set up in-plant health services, create health and safety committees, or give existing committees decision-making authority over health- and safety-related policies.

Thus the program failed to confront how the weakness and strategies of the labor movement might undermine the implementation of the act. Unions are central to the implementation of the law as it is written: they must help workers exercise their rights to participate in enforcement; they must press the DOL to develop new standards. But only a minority of the private labor force is organized, and many unions are indifferent to the problem of occupational hazards. Without public programs to help them exercise their rights, workers are unlikely to take advantage of them. But while the worker rights provisions were stronger than anyone had initially expected, they still failed to eliminate the barriers to worker action. Workers were not guaranteed compensation for participating in inspections. The law did not protect their right to refuse hazardous work. Workers were not given a role to play in the selection of company health and safety professionals. Instead, all these things remained subject to collective bargaining.

Thus the OSH Act left the existing organization of work and industrial relations essentially untouched and grafted state power onto that system. Managers remained in charge of the labor process; workers remained subordinate to them, shut out of the most important decisions about working conditions. Occupational safety and health remained a by-product of market-based investment decisions. The incentives to employers to undersupply occupational safety and health, and the disincentives to employees to participate in decisions about the conditions of work, remained in place.

[4]
The Politics of Deregulation

The OSH Act is more statist than participatory: Congress left the implementation of the rights it created to the executive branch rather than workers or unions. This guaranteed that workplace safety and health policy would be subject to intense partisan conflict, that OSHA would become the focal point for renewed opposition by employers, and that changes in working conditions would depend vitally on the changing balance of political forces.

Employers had been on the defensive during the legislative struggle over the law, and they had done a poor job of defending their interests. The passage of the act challenged them again; the threat to their property rights remained real, and the public's enthusiasm for social regulation was mounting, not waning, in the early 1970s. Although the act created a narrow range of new state powers, used imaginatively and aggressively they could easily raise the costs of production and advantage workers in the ongoing struggle on the shop floor. If workers and environmentalists forged effective alliances, they might succeed in extending the scope of regulation by building on the act.

After 1970, however, employers rose to the occasion and responded to social regulation in new and often compelling ways. The changes were dramatic.[1] There were increases in the extent of business lobbying and, more important, significant qualitative changes in how business groups lobbied. Industry created new forms of organization, overhauled its political strategy, and rehabilitated its ideology. The passage of the various health, safety, and environmental statutes in the late 1960s and early 1970s helped stimulate these changes. With OSHA, employers reacted strongly to both the costs of regulation and the assault on managerial prerogatives and criticisms of the agency were doubly fierce.

The OSH Act played an important role in helping employers rethink their understanding of the state and social reform. By bringing government into workplaces in every sector of the economy, the law encouraged firms to discover their common interests in containing and opposing regulation. In committing public policy to the advancement of new, substantive worker rights for all workers, the act also encouraged businesses to rethink their views of the positive state. In response, corporate executives reached a new level of class consciousness about reform and altered their strategies accordingly: they began to raise class issues; they offered their own positive vision of how the state might reform the market in keeping with the needs of a capitalist economy; and they developed new organizations to implement this strategy.

By the mid 1970s, this mobilization had succeeded in forcing organized labor onto the defensive. As a result, employers, not workers and unions, played a far more important role in the politics of occupational safety and health regulation, and this compounded the problems inherent in implementing the rights created by the act. In this chapter I consider how business reacted to the OSH Act, the scope and content of its political activity, and the challenge that the business offensive posed to workplace reform.

The Development of Class Organization

The mobilization of business opposition was economically rational. International markets became more competitive and profit rates declined in many industries. For many manufacturing firms, where OSHA's impact would be felt most strongly, economic conditions had deteriorated to the point where increased costs could not easily be absorbed.

The new approach was also self-conscious and carefully thought out. The legislative defeats of the late 1960s were sobering to corporate executives, who were not used to losing legislative battles; many reached the conclusion that a major overhaul in business politics was necessary. David Rockefeller provided his corporate colleagues with a particularly trenchant assessment of their situation. According to Rockefeller, the attack on business challenged the foundations of American capitalism. Critics of business, he observed, challenged the focus, aim, and scope of corporate practices. Many people had, in fact, concluded that "the system is beyond reform" and intended to "destroy the capitalist framework." The cross-industry nature of the new regulation was particularly worrisome because it encouraged people to think in global terms. "Consumerism is equated in the public mind," Rockefeller noted, "with the idea of the individual against business—*all* business" (emphasis in text). The "social contract" that bound business to society was being renegotiated, he counseled his colleagues. Unless they took a more active role in shaping it, they would see more, not less, business regulation.[2]

Others issued similar warnings. The business press filled with critical retrospectives by managers and business school professors on how and why the business community had failed to anticipate or prevent the anticorporate tide of the 1960s. Three lessons were regularly repeated: first, lobbyists had to offer positive solutions to problems; they could not simply defend the old order. Second, corporations and trade associations had to take a more active role in Washington and not lobby from distant corporate headquarters. Third, they had to teach Americans the virtues of profit making and free enterprise. In short, business had to refashion its ideology for a new age and develop a new organizational network if it was to retake the political offensive and contain social reform.

Industry began to build the necessary infrastructure in the early 1970s. The proliferation of new lobbying organizations and the revitalization of old ones was impressive. In 1972 the leaders of the largest corporations formed the Business Roundtable. It was the first new interindustry organization since the 1940s—the first to represent the largest corporations since the formation of the Business Council (BC) during the New Deal. Most important, it provided the multinationals with an organizational forum in which they could consider the principles as well as the details of basic policy issues.

Despite the elitist character of the Business Roundtable, the business mobilization was broadly based. Trade associations moved to Washington and corporate political action committees proliferated. In 1975 NAM relocated and, for the first time, registered as a political lobbying organization. Concurrently, the Chamber of Commerce stepped up its activities and sought closer ties with the academy, better relations with the media, and tighter coordination among large and small firms. The small business lobby was also revitalized. The National Federation of Independent Business, a paper organization throughout the 1960s, lobbied aggressively after 1970 and represented the special interests of small business before congressional committees overseeing the social regulatory agencies.

There were still enough different business organizations to reflect even the narrowest economic interest, but there was a new emphasis on classwide coordination. The Roundtable adopted internal policies that encouraged firms to present a common public face despite private divisions; its policy pronouncements stressed the common interests of its members. In keeping with the new spirit of cooperation, and the new awareness of overriding common interests, the Chamber of Commerce and NAM—historic rivals—attempted to merge in the mid 1970s. Although NAM pulled back at the last minute, ideology played no part in the rift. Its leaders concluded that its smaller numbers would result in the effective dissolution of the organization in the larger chamber. The Chamber of Commerce, in turn, established the Center for Small Business and the Council for Small Business in 1976 to coordinate the activities of the diverse business interests in the organization.

At the same time, business organizations made a concerted effort to mobilize and coordinate the activities of a broad spectrum of "grassroots" business interests. Corporate executives were encouraged to take a more personal role in lobbying the federal government. Parent firms were advised to expand their public affairs staffs and recruit their stockholders, employees, and home communities to lobby for and against legislation that affected them.

A wide variety of corporate-sponsored organizations emerged specifically to fight the new regulation. The Chamber of Commerce founded the National Chamber Litigation Center (NCLC) in 1977 to challenge public-interest reform and health and safety regulation in the courts and before agencies. It also organized a *Stop OSHA* campaign to coordinate business opposition to the agency in Congress. Exxon, Mobil, General Electric, IBM, Alcoa, and other corporate giants helped found the Center for Law and Economics at the University of Miami to teach corporate managers and public officials the principles of conservative economic thought. Joseph Coors, president of Adolf Coors Company, J. Robert Fluor, chair and chief executive officer of Fluor Corporation, and G. James Williams, financial vice-president of Dow Chemical, helped create the National Legal Center for the Public Interest (NLCPI), funded by the auto, steel, and oil industries. Modeling its strategy and tactics on those of the public-interest lobbies, the NLCPI sought judicial review to limit agency discretion. Its Denver affiliate, for example, helped mount the Idaho lawsuit that resulted in the Supreme Court decision that limited OSHA's right to inspect firms without a search warrant. Forty firms contributed $1 million to found the American Industrial Hygiene Council (AIHC) to fight one regulatory proposal: OSHA's generic carcinogen standard. The AIHC grew to include over 150 firms and trade associations.[3]

The Ideological Offensive

Many of the new organizations were designed to help firms mount a sophisticated challenge to health and safety regulation. The Center for Law and Economics was established to give companies "a significantly new perspective on the world." The NLCPI litigated in the name of "free enterprise" and "limited government," rather than in the interests of any single firm or industry. Michael Uhlmann, president of the NLCPI in the late 1970s, defined the organization's purpose as "the defense of the system," albeit one "whose benefits happen to be the preservation of a large degree of private decision making."[4]

Applied to OSHA, the defense of the "system" meant an effort to restore managerial prerogatives at work and reduce the costs of regulation to affected industries. Both tactics were important to a wide variety of industrial interests. Employers had failed to defend these ideas successfully in the 1960s, however, because they had treated them as self-evident propositions—component parts of the natural rights of private property holders.

In the 1970s industry shifted gears and, like health and safety reformers, sought to frame its demands with a concept of the public interest oriented to industry's needs. Many of the most important firms and interindustry groups ceased to deny the right of the state to regulate markets or the reality of the health and safety crisis. Instead, they defended their interests in more subtle ways. Most important, employers attempted to identify their particular interests in lower costs and higher profits with a general societal interest in jobs, economic growth, and capital investment. Economic growth, business suggested, was not only as important as protection but was the precondition for it. Therefore the state had to assess the effect of protective standards on the economy and take care to choose control measures that minimized the economic resources devoted to health and safety. Though subtle, this shift in approach was significant because it reopened the underlying issue: the ordering of societal priorities. Business found something to argue *for* and, thereby, the constructive approach that it had lacked; it supported "economically sound" regulation.

The climate was ripe for this kind of appeal. In the 1960s reformers could argue for the subordination of the values of production to those of protection by reminding Americans, as Secretary Wirtz had done, of what "national affluence" made possible. In the 1970s, when affluence was threatened by more competitive international markets, oil "shocks," and a host of other economic problems, appeals to the imperatives of "resource scarcity" and the importance of production took on new meaning and helped business redefine the problem of health and safety.

Concretely, employers made three related arguments: First, society should free resources for capital investment by reducing the costs of regulation to firms. In the long run, this would give Americans the best of both worlds. On the one hand, higher rates of investment would lead to higher standards of living. At the same time, living in a richer society, people would be able to take better care of themselves and work less. Second, health and safety standards should be subjected to economic review procedures that weighed the benefits of intervention to protected groups against the costs of regulation to society as a whole. Third, managers should be given discretion to organize work as they saw fit. If the state eschewed bureaucratic "interference" with production and allowed firms to adopt the most efficient protective measures, it would encourage industrial innovation and productivity growth.

By far the most important component of the business ideological offensive was its claim that society's interest in economic growth and capital investment was equal to, if not prior to, its interest in protection. As I argued earlier, economism is reproduced, in part, because workers and labor organizations believe that their jobs and standards of living depend on private investment. If they act in this way, they are likely to limit their demands to things that are compatible with corporate profitability. Public policy that internalizes this point of view is likely to reinforce this kind of worker activity.

Industry began to demand economic relief from health and safety standards as soon as OSHA started regulating. Nonetheless, it took several years for business groups to figure out how to make this claim generalizable. At first, individual firms and trade associations asserted the needs of particular industries and challenged the feasibility of particular standards. The plastics industry's opposition to OSHA's vinyl chloride rule illustrates the basic strategy.

In 1974 several companies in the vinyl chloride industry, which employed 7000 workers, reported that 16 employees who worked in polyvinyl chloride production plants had died from a rare form of liver cancer. The industry is divided into three segments: the production of vinyl chloride monomer (VCM), used in the manufacture of polyvinyl chloride (PVC); PVC resin production; and the fabrication of consumer products from PVC, such as pipes, tubing, floor tiles, and phonograph records. Since the 1920s, research has indicated that workers in all three segments are exposed to abnormally high risks of liver and kidney damage, as well as skin, stomach, and circulatory disorders. In response, one company, Dow Chemical, voluntarily reduced exposure levels to 50 ppm on an eight-hour time-weighted average and set a maximum level of 100 ppm. But the rest of the industry rejected claims that vinyl chloride was toxic and adopted the

ACGIH 500 ppm threshold limit value (TLV). They maintained it despite growing evidence that VCM was also a carcinogen.[5]

In 1971 OSHA adopted the 500-ppm exposure level when it adopted the ACGIH's list of consensus TLVs. After the 1974 reports, it proposed to reduce the permissible exposure to "no detectable level"; the polyvinyl chloride industry strongly objected to the proposal. Although the industry challenged the evidence relating worker exposure to cancer and the technical feasibility of achieving this level of protection, it stressed the economic effects of the standard. The Society of the Plastics Industry (SPI), representing the affected firms, issued a report that argued that a no-detectable standard would close down all PVC resin plants and hurt all industries that used vinyl chloride. In total, $65 billion in production and 1.6 million jobs would be lost. The results, according to the SPI, would be "catastrophic."[6]

Although this strategy remained attractive to many firms and industries, the courts forced employers to find an alternative approach in 1974 with a court of appeals ruling on OSHA's asbestos standard. The health hazards of asbestos were well known when OSHA was established; they had played an important part in the congressional hearings on the OSH Act. As a result, the agency acted quickly and issued a two-fiber standard as its first new permanent rule in 1972. But OSHA concluded that compliance was not immediately feasible for the industry and, based on economic and technical considerations, allowed firms four years to comply. In response, the AFL–CIO's Industrial Union Department challenged OSHA's standard, charging that its interpretation of Sec. 6b(5)'s reference was incorrect. That section, it maintained, precluded the agency from taking economic factors into account in health standard setting.[7]

The court of appeals rejected the federation's argument and upheld an economic reading of the Sec. 6b(5) reference to feasibility, but its decision also blocked the efforts by particular industries to seek relief because of high compliance costs. In *IUD* v. *Hodgson*, the court ruled that OSHA standards could put individual firms out of business and cut into the profits of all firms in a particular line of business. A standard was economically infeasible only when it threatened the existence of an entire industry. According to the court, common usage suggested that "a standard that is prohibitively expensive is not 'feasible.'" However, the court cautioned,

This qualification is not intended to provide a route by which recalcitrant employers or industries may avoid the reforms contemplated by the Act. Standards may be economically feasible even though, from the standpoint of employers, they are financially burdensome and affect profit margins adversely. Nor does the concept of economic feasibility necessarily guarantee the continued existence of individual employers.[8]

Once the court of appeals made clear that the courts would not look favorably on efforts by industries that pleaded special cases, trade associations and other business groups shifted their strategies and asserted the value of production in general. The standards of OSHA, they argued, threatened the viability of the economy as a whole. The SPI's challenge to the vinyl chloride standard foreshadowed this strategy by estimating the economic impact of the standard on over-all production. The court of appeals also seemed to approve of this direction in *IUD* v. *Hodgson* when it noted that "complex elements" were relevant to the determination of economic infeasibility, including the impact of standards on the competitive structure of American industry or the ability of an industry to compete in the world market. After 1974, these themes came to dominate industry's case against social regulation.

To be sure, individual firms and industries continued to challenge particular standards and argue that the costs of compliance were too large to bear, or that mandated controls were technically infeasible, thereby hoping to satisfy the criteria established in the 1974 decision. In 1978 the cotton textile, lead, and chemical industries challenged OSHA standards on the grounds that they could not absorb the capital costs of compliance and "whole product lines" would go out of business.[9]

After 1974, however, firms and industries stressed society's general economic interest as much if not more than their own particular costs. Regulation meant lost productivity, price inflation, and declining international competitiveness. On this level, business groups spoke with one voice, albeit in different tones. The Chamber of Commerce struck a populist note and portrayed the problem in classically anti-statist terms.

All of us pay for OSHA's failures. We pay as consumers when the goods we buy cost more in the marketplace. We pay as taxpayers with more and more whittled from our paychecks to fund an agency that is heavy on expenses but lean on results.[10]

The Business Roundtable's 1979 *Cost of Government Regulation Study* took a more measured tone and appealed to the norms of efficiency. But it reached the same conclusion. The "imposition of large cost burdens on the private sector" rested "ultimately on the U.S. economy." The study estimated that the "incremental costs" to 48 member firms—costs in addition to what firms would have undertaken in the absence of regulation—of 6 regulatory agencies, including OSHA and EPA, were $2.6 billion in 1977, or more than 10% of their total capital expenditures in that year.[11] Moreover, the Roundtable argued, these costs were only a small part of the total burden. There were "many less visible secondary effects that cause substantial incremental costs . . . to society generally," including "losses in productivity of labor, equipment and capital, delays in construction of new plants and equipment, misallocation of resources and lost opportunities."[12]

If government was to take these considerations into account, it had to have a way of assessing the economic implications of agency standards. Neither the OSH Act nor judicial review of OSHA rules provided a formal mechanism to weigh the impact of standards on the economy as a whole. Accordingly, business groups urged Congress and the executive to adopt an economic review procedure in which "objective" third parties would use "neutral" decision-making tools to review agency rules.

The AIHC recommended that the assessment of health hazards "be taken out of the government arena" and viewed outside the "political framework" by "the best scientists government can hire."[13] Basic social regulatory policy would begin from the premise that there are "socially acceptable" levels of risk. These levels, the AIHC suggested, would be established by panels of independent experts who took economic as well as scientific factors into account.[14]

Cost-effectiveness and cost–benefit tests were the preferred ways of making sure that society's general interest in economic growth and capital investment was respected. In general terms, cost-effectiveness tests require that agencies achieve their stated goals in the most efficient manner possible. Cost-benefit tests apply a stricter standard and require agencies to forgo regulating in cases in which the net costs of a particular action outweigh the net benefits. By the late 1970s, a consensus had formed among business groups that both tests had to be applied to agency rulemaking. Where health and safety agencies resisted these procedures, business groups also argued for centralized oversight that imposed economic review on regulators.

The defense of managerial prerogatives, a long-standing priority of industry, was also rehabilitated by this emphasis on economic growth and investment. In defending the freedom of managers, employers suggested three major changes in occupational safety and health regulation: the adoption of performance standards, "cooperative" enforcement, and limits on worker rights to participate in the determination of working conditions. Taken together, these steps, they claimed, would also help revitalize the economy.

Performance standards establish hazard-reduction goals rather than specify changes in physical plant, machinery, or work practices. Employers argued for them on several grounds. Standards that mandate detailed design changes, they claimed, were counterproductive. If OSHA set a goal of a 10% reduction in work injuries, firms could use the expertise of plant engineers and supervisors to meet it in ways that distant Washington bureaucrats, inexperienced in production, were unlikely to discover. This reform was particularly necessary, they argued, because productive techniques and compliance methods changed constantly; detailed regulations might force firms to adopt outdated rules. Moreover, engineering controls often increased costs far beyond what was necessary and at the same time "stifled innovation." Firms could accomplish the same protective goals if they were allowed to substitute personal protective devices such as ear plugs and respirators for the engineering controls generally relied on by OSHA.[15]

Employers also argued that OSHA should cooperate with, rather than "punish," firms that failed to provide healthy and safe workplaces. Penalties should be deemphasized, and OSHA inspectors should consult with firms about ways to improve their health and safety practices. Penalty-based enforcement was both inefficient and overly "adversarial." According to business lobbyists, most firms wanted to comply with OSHA regulation, but many were unaware of the complex and often confusing rules adopted by the agency. In this context, the "polarizing" presence of inspectors eager to find violations and levy fines increased employer hostility to the entire regulatory effort.

As an alternative, employer representatives suggested that OSHA negotiate compliance agreements with firms and that firms with good safety records be granted blanket exemptions from OSHA inspections. Ideally, OSHA should completely forgo penalties except in cases of repeated violations; should allow the agency's regional directors to adjust citations and penalties at informal conferences requested by employers; and should discipline "antibusiness" compliance officers—those whose citations were systematically overturned by the Occupational Safety and Health Review Commission.[16]

Industry also argued that Congress and the agency were misguided in attempting to involve workers in the implementation of the act. These efforts increased labor–management conflict and complicated OSHA enforcement because worker participation encouraged employees and unions to use their health and safety rights as weapons in collective bargaining. According to the Chamber of Commerce, OSHA's "interjecting itself into the collective bargaining setting" had "a potential for seismic repercussions." It was "an opportunity for abuse and union harassment of employers." Also, by making occupational safety and health adversarial, state-mandated worker participation undermined the contribution that in-plant programs and voluntary management committees made to workplace health and safety. Corporations were interested in worker protection, and OSHA could supplement their efforts by educating and training employers and employees. But aggressive enforcement of worker rights to participate in implementation of the act, including compensation for time spent with inspectors, protection from retribution for refusing hazardous work, and interference with company pay scales and hiring practices in the interests of worker safety, alienated employers and discouraged private efforts to improve working conditions.[17]

Many of these claims had been made in defense of the existing system in the late 1960s; framed in this new way, however, they took on new meaning. No longer were employers defending state agencies and private professional groups. Rather, they were championing the rights of society to higher standards of living. To be sure, their own interests would also be promoted by deregulation. But this was a by-product of reforms necessary to promote everyone's interest in the health of the economy.

The Rehabilitation of Market Capitalism

Having shifted from opposition based on the costs of particular standards to support for economic review of the effects of rules on the macroeconomy, employers were better positioned to answer health and safety reformers' claims about the failures of market capitalism. Reformers had argued that markets failed to provide an adequate level of health and safety and that society's general interest in protection required that government impose regulations that increased the firms' costs of production. In contrast, industry's new view of the problem argued that by inhibiting growth, regulation caused another kind of market failure. Moreover, regulation promoted workers' particular interests in protection against society's general interest in capital investment.

The significance of this ideological redefinition cannot be overemphasized. As formulated, it relegitimated market capitalism. Business had fought social regulation in the 1960s by defending the free market, and health and safety reformers had been able to use the classic theoretical defense of the market system—microeconomic theory—against them. According to microeconomic principles, when markets "failed," that is, when all the costs (and benefits) of production were not reflected in the prices of goods and services, government had a positive obligation to make sure that the "external" costs (and benefits) to third parties such as workers and consumers were "internalized." Whatever the merits of the microeconomic approach, the reformers' claims, posed in this language, were hard for employers to rebut.

By shifting attention to the macroeconomy, this new view undercut the conventional market-failure argument. Viewed from the perspective of the growth of the system as a whole, social regulation imperiled rather than perfected the market system. The macroeconomic critique also restated the logic of economism. Because workers were dependent on capital investment for jobs and income, the interests of business in profits were general interests and took precedence over all other interests.

Still, once employers had translated their opposition to OSHA into a generalizable economic critique, a number of rather difficult political issues remained to be resolved. Almost all business groups agreed that OSHA should be required to do some form of cost–benefit analysis: the agency should be required to measure the aggregate economic costs of regulation and justify them with reference to specific and quantifiable benefits. But there were four potential pitfalls in this approach.

First, most interested parties recognized that it was difficult to estimate the economic value of health benefits, including lives saved. Many regulatory reformers who were otherwise sympathetic to the economic critique of social regulation were concerned that overly rigid cost–benefit tests would artificially bias policymakers against high levels of protection. Their concerns had to be addressed.

Second, given the right assumptions, many costly regulations could pass cost–benefit tests. The methodology is notoriously subjective; the analyst must make several judgments about what to count as costs and benefits and how to monetize them. Consider the problem of valuing a life, for example. Policymakers have used estimates varying from $200,000 to $7 million.[18] They also have had to consider how the value of future lives should be calculated. Should lives saved 20 years from now be discounted, as economists and accountants routinely discount income streams to assess the present value of assets? If so, at what rate? A 10% discount rate, a convention in economic forecasting, all but eliminates the value of long-term health benefits. In contrast, if future benefits are not discounted, the cumulation of lives saved in the future can lead to extremely high benefit estimates.

Third, many widespread hazards are extremely costly to workers and taxpayers. If the analyst chooses to recognize and monetize the entire range of direct and indirect effects of injuries and disease, including lost income, medical care, job retraining, lost productivity, public assistance, and pain and suffering, the benefits of regulation are enormous. If the analyst does not discount these benefits and uses liberal estimates of the value of lives saved, almost any standard can be justified.

A cost–benefit study of the 1972 asbestos standard, for example, performed independently of OSHA and presented at a 1975 DOL conference, demonstrated the extreme sensitivity of cost–benefit test results to the analyst's assumptions. Employing a wide range of reasonable-alternative assumptions, Russell Settle came up with 72 different estimates of the net benefits of OSHA's two-fiber standard. The benefit–cost ratios of these estimates ranged from .07 to 27.70.[19]

This variability did, in fact, discourage many regulators who were otherwise cost conscious. Dr. Morton Corn, President Ford's choice to be OSHA's third assistant secretary, described his experience with the methodology in this way:

> After arriving at OSHA, I engaged in an in-depth consideration of cost–benefit analysis, applying the methodology to the coke-oven standard. . . . With the dose-response data at our disposal, various assumptions were used to ring in changes on different methodologies for estimating benefits. The range in values arrived at, based on the different assumptions, was so wide as to be virtually useless. The conclusion I reached after this exercise was that the methodology of cost-benefit analysis for disease and death effects is very preliminary, and one can almost derive any desired answer.[20]

Finally, the creation of an economic review process raised serious jurisdictional issues. Administrative regulation involves all three branches of government in a complex and mutually interdependent process of rulemaking. The boundaries between their respective jurisdictions have never been clear, and all jealously guard their prerogatives to supervise the bureaucracy. Congress insists on the sanctity of its legislative mandates; the executive branch argues for agency discretion; and the courts maintain the final right to determine the legitimacy of agency decisions.

The economic review process changes this division of authority. A hard-and-fast cost–benefit test imposed by Congress limits agency discretion. The same test, imposed by the White House on the agencies, interferes with congressional intent in authorizing legislation. And challenges to the economic review process force the courts to make complex value judgments about technical as well as political issues.

These problems were not insurmountable, as the Business Round-table's carefully considered 1980 recommendations on regulatory reform demonstrated. The Roundtable accepted the fact that automatic decision rules, like cost–benefit tests, were unrealistic and would probably be unacceptable to the White House, which had a vested interest in maintaining its discretion over agency policies. Instead, it proposed the following:

■ All agencies, including independent commissions, should be required to compare the costs of alternative approaches to problems and justify the selection of more costly alternatives.

■ Agency regulatory analyses should be part of the rulemaking record and subject to judicial consideration when agency rules were subject to final judicial review.

■ White House oversight of the review process should be centralized in a single agency or body appointed by the president and subject to confirmation by the Senate.

■ The courts should be ordered to construe congressional delegation of authority narrowly unless statutes contained a clear statement of authority.

In sum, the Roundtable proposed that agencies be forced to do cost–benefit tests but that final decisions about protection would be made by the White House and scrutinized by the courts.[21] The Chamber of Commerce and NAM quickly rallied around similar proposals.

Allies in the Academy

The success of the business offensive depended in large part on the employers' ability to redefine the general and particular interests at stake in social regulation and sell those ideas to the wider society. With the economic resources at its disposal, the business lobby mounted a concerted campaign to shape elite and mass opinion, using advocacy advertising to reach lay audiences and financial support for conservative foundations and think tanks, such as the American Enterprise Institute for Public Policy and the Hoover Institute, to influence academics.

Many academics did, in fact, help business make its case. Economists were particularly active in this movement. Political scientists and policy analysts also played critical roles by producing distorted economic impact studies that demonstrated the high costs of regulation, one-sided critiques of conventional forms of standard setting and enforcement, and theoretical briefs for cost-benefit analysis and other forms of economic review.

Dr. Murray Weidenbaum's efforts are probably best known. His estimate that federal regulation cost Americans approximately $66 billion in 1976 and grew to over $100 billion in 1979 appears to be the most often cited impact study. President Reagan used Weidenbaum's figures in his 1981 message on the economy, and they are regularly repeated in corporate advertising.[22]

Working under the auspices of the Center for the Study of American Business at the University of Washington at St. Louis, Weidenbaum and his associates came up with a fairly simple way of tackling the difficult problem of estimating the total costs of regulation. First, they collected whatever data were available, including company reports, to calculate the direct "costs of compliance" in 1976. These were $63 billion. Then they estimated the administrative costs of regulation by adding the budgets of the various agencies and departments involved in regulation. These costs were $3.2 billion. Thus the total costs of regulation were about $66 billion. Of this total, occupational safety and health accounted for $4.5 billion, including $483 million in administrative costs and $4.02 billion in compliance costs.

Using these figures, Weidenbaum calculated a *multiplier*, or the ratio of compliance costs to administrative costs. This multiplier—20 in 1976—was then used to estimate total regulatory costs in subsequent years. Weidenbaum's 1979 estimate of a total regulatory burden of more than $100 billion was based on administrative costs of $4.8 billion and estimated compliance costs of $96 billion (20 multiplied by $4.8 billion). The total costs of regulation in 1979 amounted to 4.3% of the GNP. Weidenbaum actually believed that these figures underestimated total regulatory costs because they did not take the "induced" effects, such as the impact of regulation on labor productivity or innovation, into account.[23] Nonetheless, they suggested the size of the burden and the urgency of the crisis.

As many critics have pointed out, there are a host of flaws in Weidenbaum's study, ranging from double counting to conceptual confusion about the nature of economic costs. The list is rather long, but four errors are particularly egregious. First, by using a constant multiplier from one year to another, the study ignores the fact that the costs of regulation are usually high in the initial years and then diminish as firms come into compliance. Second, Weidenbaum failed to distinguish between social regulation and the ordinary operations of government, including, for example, the costs of Internal Revenue Service filings. Third, the study did not acknowledge the difference between *incremental costs*, that is, costs due to regulation, and expenses that firms would undertake on their own. Fourth, Weidenbaum failed to distinguish between regulatory programs that transfer costs from one party to another—clean-air programs, for example, reduce property damage caused by pollution but raise industries' production costs and consumer prices—from programs that simply increase total costs to society.[24]

For Weidenbaum and his supporters, however, the numbers were less important than the underlying ideas. Weidenbaum, later appointed chair of the Council of Economic Advisers by Ronald Reagan, was adamant about the procapitalist implications of his study and unimpressed by his critics. His work was intended, he told Congress in 1979, "to shift the public dialogue onto higher ground." From that ground, environmentalists, consumer groups, and other reformers could be seen for what they were, "self-styled representatives of the public interest who have succeeded . . . in identifying their personal prejudices with national well being." In contrast, corporations "serve the unappreciated and involuntary role of proxy for the overall consumer interest."[25] Weidenbaum was willing to admit that his figures could be challenged, but precise figures were beside the point. If anything, his figures underestimated the real impact of regulation because its costs were "immeasureable"—regulation threatened the "basic entrepreneurial nature of the private enterprise system."[26]

Though popular with politicians and the media, Weidenbaum's study was sufficiently flawed to leave the defense of economic review vulnerable to its critics. But other, more careful policy analysts provided a more rigorous critique of the agencies' basic approach to rulemaking, based on two related ideas. First, market exchanges remained the appropriate point of departure for designing health and safety policy. Second, social regulatory policy failed to acknowledge the inevitability of risks and the desirability of risk taking in modern industrial societies.

The first point, that regulatory programs should be evaluated according to the norms of the marketplace, proved increasingly popular as conservative economic ideas resurfaced in the wake of the 1973–1975 recession. Echoing business complaints that the heavy hand of government had stifled economic growth and innovation, many policy analysts rediscovered the case for markets. In a highly influential critique of OSHA written for Congress, Richard Zeckhauser and Albert Nichols, both policy analysts at the Kennedy School of Government at Harvard University, suggested that the norms of efficiency required that the agency rely more heavily on market forces. Conventional standard-setting and enforcement programs were inflexible, they claimed. At a minimum, OSHA should use performance standards; but economic incentives systems, such as workers' compensation and labor markets, were even more desirable.[27]

Two related American Enterprise Institute (AEI) studies argued for a complete overhaul of the regulatory apparatus and a return to a market-based system supplemented by a reformed workers' compensation program. In the first study, Robert Smith argued that "the safety and health mandate of the Occupational Safety and Health Act of 1970 is inconsistent with the goal of promoting the general welfare" because it "force[s] more safety and health on society than workers would choose for themselves if they had to pay the costs of safety and health directly."[28] In a companion piece on workers' compensation, James Chelius argued that the OSH Act "scapegoats" employers who, in fact, play only a small role in creating occupational hazards. According to Chelius, the most efficient program would allow workers to choose between jobs with different wage and risk characteristics. Government regulation of labor markets was justified only when workers did not know about hazards at work or could not bargain for risk premiums for assuming them. Where labor markets are competitive, and workers are informed about the risks they face, regulation is unnecessary.[29]

Most academic critics of OSHA recognized that labor markets and workers' compensation were often inappropriate ways of protecting workers against health hazards. For these dangers, they recommended the kinds of economic review procedures that business had endorsed. Zeckhauser and Nichols, for example, recommended that OSHA should be forced to take "explicit consideration of economic costs" when choosing health targets. This would lead OSHA to focus its standard setting on areas where it could achieve "the greatest health gains for whatever resource costs they entail."[30] Recognizing that this approach to standard setting contradicted the act's provisions, they recommended amending the law to make economic review possible.[31]

Cost-effectiveness and cost-benefit tests appealed to policy analysts because they introduced efficiency criteria in government standard setting. To nearly all policy analysts trained in neoclassical economics, sound policy is efficient in a particular sense: when scarce resources are devoted to their most economically productive uses. This is accomplished when certain marginal equalities are satisfied. For example, firms should produce a particular good until the marginal revenues from its sale equals the marginal costs of its production. Similarly, employees should work up to the point where the marginal benefits of their labor (i.e., wage gains) equal the marginal costs (i.e., lost leisure time).

The neoclassical view also argues for using market values to determine the costs and benefits of state action. Thus economic reviewers should, wherever possible, calculate costs and benefits by using the prices that individuals place on them when they act in markets. For example, the value of health and safety at work should be determined by observing the tradeoffs that workers themselves make between wages and safety. Costs, in turn, should be calculated by adding the market prices of the resources consumed, and opportunities forgone, as a result of state action. By following these rules, policymakers will choose policies that are "optimal" from an economic point of view.

W. Kip Viscusi, a professor of business administration at Duke University and a consultant to OSHA during the Reagan administration, states the case for both of these criteria in his suggestions for reforming OSHA rulemaking:

> First, the government should select the policy that provides the greatest excess of benefits over its costs and, since one alternative is to do nothing, it should not adopt any policy whose costs exceed its benefits. Second, to obtain the highest net gains from policies, the scale of the programs should be set at levels where the incremental benefits

just equal the incremental costs; further expansion or reduction in the policy will produce lower net benefits overall. Third, all policies should be cost-effective, that is, the cost imposed per unit of benefit should not be greater than for other policies.[32]

Finally, wherever possible, administrators should use workers' "willingness to pay" as revealed in risk premiums to value the benefits of standards. If risk premiums do not provide reliable information, policymakers should rely on other measures of how workers value protection, including survey research, to calculate benefits.

Conservative academics helped buttress industry's claim that risks were an inevitable by-product of industrial society. Indeed, Aaron Wildavsky, head of the School of Public Policy at the University of California at Berkeley, went one step further and argued that risk taking in markets was likely to provide more health and safety than protective regulation did. Wildavsky's argument was simple and, on its face, compelling. "In the hundred years from 1870 to 1970," he observed, "every increase in industrialization and wealth, except possibly at the highest levels, was accompanied by a corresponding increase in safety from accident and disease." "Richer is safer," he claimed, and he used the case of workplace safety to illustrate the point. According to Wildavsky, the simplest way to reduce workplace accidents is to improve machinery, reduce the number of workers involved in production, and shorten the workday. Economic growth and technical change accomplish all these things. Conversely, by reducing productivity and inhibiting innovation, protective regulation discourages them.[33]

Wildavsky also suggested that, divorced from the discipline of the market, and taken to its logical conclusion, the ethic of protection would lead to demands for unlimited risk reduction.

If they are permitted to proliferate, direct demands for reduction of risk group-by-group, case-by-case, are inexorable. For one thing, the politics of anticipation requires that all possible sources of risk be eliminated or mitigated. Since these sources are virtually infinite in number, subject only to the fertility of the imagination, there is no limit on what

can be spent on them. For another, there is no principled
reason why risks that affect certain groups should be
reduced while those potentially affecting other groups are not.[34]

The political system would compound the problem because "accommodation by logrolling will lead to the usual coalitions of minorities,"
each assenting to the other's demands for protective legislation.
Echoing the economists' critique of departures from the market allocation of goods and services, Wildavsky concluded that "the result
will be more 'safety' than anyone would choose to buy."[35]

The Perspective Shifts

In many respects, these claims are traditional. Defenders of the status
quo in capitalist democracies have generally resorted to two kinds of
arguments. The first asserts that capitalism as an economic system is
superior to any other possible system; the second asserts that markets
promote individual liberty. Employers and their allies in the academy
attacked OSHA and social regulation on similar grounds. Protective
regulation, they argued, inhibited the productive efficiency of the
economy and the freedom of the market.

But employers had learned how to make this case under new conditions; new hazards, new movements, and new values forced them
to rehabilitate their ideology and organizations. Few people were
convinced of the inherent virtues of self-interested action in markets;
people wanted to believe that they were pursuing the public interest.
Business learned to make this case: by serving society's general interest in capital investment and economic growth, business too served
the public interest. Any measure that increased its ability to innovate
and compete could be defended in these terms. In this way, business
retook the ideological offensive.

[5]
Labor's Defense of
Social Regulation

he OSH Act threatened employers
and forced them to defend their
property rights, but it offered the
labor movement an opportunity to
take the offensive on three fronts:
to argue for an expanded vision
of the rights of workers, to link
workers' demands for protection
to a general program of progressive social reform, and to establish itself as a leader among the
new social movements that developed in the 1960s
and in the early 1970s.

To do these things, the unions had to change the way they
approached work, the environment, and the state. As Chapters
2 and 3 indicated, economism continued to dominate organized labor's approach to social change throughout the 1960s.
Barring a few exceptional cases, the union leadership had not
been drawn to the quality-of-life agenda or demands for a more
participatory society and economy. A complex set of conjunctural
factors, rather than a major change in organized labor's political and
economic strategies, had made the passage of the OSH Act possible.

For the most part, this remained true after 1970. As this chapter describes, despite some signs of change on the part of rank-and-file workers and some unions, and despite efforts by the environmental movement to forge stronger links with organized labor, the patterns established in the postwar accord held.

The Possibilities

The act provided a stepping-off point for organized labor by giving employees a new kind of right. The right to healthy and safe work is, as I have stated, a dramatic extension of the way that worker rights have been understood in the welfare state: it is substantive rather than procedural, and it draws the state into supervising managerial decisions over the actual conditions of work. In giving workers the right to participate in OSHA enforcement, and empowering the state to regulate the workplace, the act suggested broader democratic principles that led naturally to other, equally radical rights: that workers should influence the local conditions of work and that, through the state, the public might exercise greater social control over the processes of capital investment and technological change.

These broader democratic principles are particularly important to a movement that seeks to unify people in diverse economic and social settings. Not all workers, after all, are exposed to the same levels and kinds of risks. Blue-collar workers generally have more job-related injuries than white-collar workers. The injury rate of laborers is more than two-and-one-half times that of craft workers and four times that of service workers (see Table 5.1). By linking the right to occupational safety and health to the idea of state regulation of production and investment, the labor movement could overcome these divisions by appealing to a shared interest in public control over the economy, including job creation, plant location, and community development.

The vision underlying the OSH Act could also be used to legitimate rights to information about technology, investment, and employer practices. Workers could demand access to medical records, information about management decisions affecting health and safety,

Table 5.1.   Job Risk by Occupation

Occupation	Ratio Index
All occupations	1.00
Laborers	3.70
Transport equipment operatives	2.09
Operatives	1.79
Craft workers	1.40
Service workers	.92
Managers and administrators	.28
Salesworkers	.28
Clerical workers	.24
Professional, technical, and kindred workers	.21

Note: Excludes data for agriculture, forestry, and fisheries; private households; and the public sector. Indexes are derived by dividing the occupation's percentage of total injuries by its percentage of total employment.
Source: Norman Root and Deborah Sebastian, "BLS Develops Measure of Job Risk by Occupation," *Monthly Labor Review* 104, no. 10 (1981), table 1, p. 28.

prenotification about the introduction of new production methods, and independent surveys of the potential health impact of changes in the workplace. Workers could also demand new rights to their jobs: employees incapacitated by health hazards at work could be given the right to other, equally well-paid work; firms could be required to protect employees from shutdowns caused by high abatement costs.

In addition, the act's right to safe work could be used as a first step toward redefining the very notion of occupational health and turning criticisms of the impact of industrial capitalism on the quality of life into a positive program for new forms of production that promoted noneconomic values, including emotional well-being and job satisfaction. The growing interest in stress and other psychosocial health hazards caused by the work environment points in this direction. The Scandinavian countries have done pioneering research on the subject.[1] This is an obvious next step for a society in transition from conventional forms of industrialism to newer, automated technologies, including information processing, which promise to increase stress at work and, consequently, increase the incidence of diseases such as hypertension.

Occupational safety and health reform also provided the labor movement with a way of forging closer links with nonlabor groups and, at the same time, making good on its claim to represent workers as a whole rather than only union members. Throughout the 1970s and early 1980s, public opinion supported social regulation. Despite long-standing opposition by the majority of Americans to greater regulation of business in general, opinion polls indicated that the public clearly endorsed health, safety, and environmental reform. People were skeptical about the benefits of economic regulation of industries such as airlines and trucking, but they strongly supported protective legislation designed to improve product safety and air and water quality, and reduce workplace hazards.[2]

According to survey research, Americans were worried about health and safety hazards and wanted the government to take strong action to protect them. One poll found that two-thirds of its respondents supported the idea that regulators set standards that lowered exposure to carcinogens to "zero" or the "lowest possible level."[3] In another, 85% endorsed the idea that the federal government should set increasingly strict standards for auto emissions.[4] An ABC News/Harris survey reported that 93% of those polled believed that federal standards prohibiting dumping of toxic chemicals should be stricter;[5] 86% wanted the provisions of the Clean Air Act maintained or strengthened; and 93% felt this way about the Clean Water Act.[6]

Most people also believed that work was dangerous, that corporations would not take measures on their own to protect workers, and that OSHA should not be deregulated. Pollsters found that a majority of Americans believed that exposure on the job was an important source of cancer but that only one-third thought that industry was trying to reduce exposure to cancer-causing agents.[7] A CBS/New York Times survey found that nearly three-quarters of those polled believed that government should set standards and implement them.[8] In a 1981 Harris survey, taken at the height of Reagan's deregulation campaign, two-thirds of those surveyed rejected "cutting back sharply on the enforcement of employee safety regulation by OSHA."[9]

Table 5.2. Are the Costs of Regulation Worth It?

	Percentage answering yes	Percentage answering no
1. Social regulation reduces capital investment in plant expansion and modernization?	51	37
2. Federal regulations add to the cost of consumer goods?	69	18
3. Are the costs of regulations worth it: to protect workers' health and safety?	52	12
to ensure safety/dependability of products or services?	47	15
to ensure equal employment opportunities?	42	21
to protect the environment?	42	19

Note: Totals do not include those with no opinion. Question wording is as follows: 1. "The money that business spends meeting government requirements for environmental protection, health and safety, equal employment opportunities, etc., has significantly reduced the amount that business can invest in the expansion and modernization of plants and equipment." 2. "Federal regulations and requirements add to cost of consumer products or services." 3. "Costs added by regulations are worth it to . . . ."
Source: Opinion Research Corporation, *Government Regulation of Business: The New Outlook, the New Realities* (Princeton, N.J.: Opinion Research Corporation, 1978), pp. 34–36.

Majorities understood that protective regulation was costly and were prepared to pay the price. A 1982 Harris survey on consumerism for Atlantic Richfield Corporation found that people thought the costs of consumer protection were worth the effort; there was "hardly any support at all . . . for regulatory rollback or dismantling in the consumer protection area."[10] A survey done for the Continental Group, a company with interests in timber, oil, gas, and land development, found that half of the respondents were willing to accept a slower growth rate to protect the environment. Only one-quarter wanted to lower environmental standards to achieve economic growth; the rest believed that both growth and protection were possible.[11] The findings of a 1978 survey, displayed in Table 5.2, demonstrate the public's commitment to protection, including occupational safety and health regulation, despite the costs.

Perhaps most important, public opinion polls suggest that the support for protective regulation was broadly based and provided the labor movement with an issue that could widen its appeal. The majorities in favor of product safety, for example, encompassed the traditional working-class and minority constituencies of the New Deal coalition, as well as many in the middle class. The breadth of the potential coalition was enormous, ranging from poor blacks to white suburbanites.[12] The overwhelming majorities in favor of air- and water-pollution control suggest the same conclusion. In all these cases, only the poorest, rural residents, and those with the least education balked at strict regulation. Even among them, the differences between their attitudes and those of the general population were not large. In short, the public was ready to see worker protection as part of the public interest rather than a narrow union demand. If the labor movement framed it in these terms, it could strengthen its claim to speak for the common good.

The act itself provided several natural bridges between labor and other social movements. By emphasizing health hazards, for example, it facilitated links between the unions and the environmental, black, and feminist movements. The noneconomic character of workplace reform appealed to environmentalists, who already saw the workplace as a microcosm of the larger environment, one that threatened workers inside plants and residents in surrounding communities. Many were eager to forge links between their organizations and the union movement to force the implementation of the whole range of new health, safety, and environmental statutes.

Occupational health is a particularly important issue to blacks and women. Both groups are employed in large numbers in high-hazard industries, but both find it difficult to improve their situation. Blacks have particularly high morbidity rates from hypertension and other stress-related disorders. Black steelworkers, for example, suffer disproportionately from hypertension and arteriosclerotic heart diseases compared to the general population or coworkers. Compared to white coworkers, blacks in laundry and dry cleaning firms have twice the death rate from circulatory diseases. Often, discriminatory employment practices contribute to these conditions by forcing black workers into the least desirable jobs. But industry and the labor movement have ignored the problem. Indeed, the special health hazards of black workers have been all but invisible.[13]

Stress is also a major occupational hazard for employed women. Overall, women's safety record is better than that of men, probably because relatively few are in hazardous blue-collar manufacturing jobs. But women's dual role as paid worker and unpaid homemaker increases the psychological and physiological costs of labor-force participation. "Pink-collar" clerical jobs are characterized by factors most clearly associated with stress, including machine-paced work, low compensation, excessive hours, boredom, and discrepancies between education and occupational status. The double burden of wage work and homework, especially for the growing number of single women parents, compounds these problems. But, again, women's health hazards have traditionally received little attention.[14]

Worker rights to participate in OSHA enforcement and to know about hazards provided a natural bridge to the public-interest movement as well. They appealed to Ralph Nader during the legislative struggle and remained important issues after the passage of the act. Although often indifferent and even hostile to union wage demands, middle-class reformers were comfortable with the participatory spirit behind demands for information about working conditions and agency actions. As an administrative agency, OSHA was also an obvious target for public-interest groups that sought to play the role of watchdog and hold the regulatory bureaucracy accountable.

Thus the OSH Act, and business opposition to it, posed a dual challenge to organized labor. First, the implementation of the act would require unions to force Congress and the White House to implement a law opposed by employers. If the unions were to accomplish this, they would have to forge the broadest possible coalition of liberal and progressive groups and convince the public and public officials that occupational safety and health was a general societal issue rather than a demand, by labor, for "special" organizational or economic advantages.[15] Second, to expand the act's rights, the labor movement would have to demand further reforms from employers and government. Managers would have to accept health and safety committees with rights to inspect plants, shut down machinery in imminent-danger situations, veto changes in work organization, and inspect new equipment before its introduction. The state, in turn, would have to mandate many of these provisions, force employers to increase their financial commitments to in-plant health services such as occupational clinics, and undertake a more extensive research effort into the causes and consequences of occupational diseases.

These two challenges were related because they depended on the same strategic shift. To secure the implementation of the act and further reforms, the labor movement would have to mobilize a coalition of unions, environmentalists, public-interest groups, civil rights groups, and feminists to pressure Congress and the White House. But political mobilization meant that workplace safety and health had to be cast in the broadest possible terms. In short, to make occupational safety and health reform effective, organized labor had to fold it into a broad conception of participatory democracy and social reform.

Signs of Change

There were some significant signs of change. Survey research indicates, for example, that rank-and-file workers became more aware of occupational hazards after 1970 and sensitive to health hazards in particular. While official injury rates declined after 1969, the number of "work-related injuries" reported by workers on the University of Michigan Quality of Employment Surveys (QES) increased between 1969 and 1972. Workers also began to focus on occupational diseases: the ratio of health to "traumatic" or safety injuries in the total number of worker-reported injuries increased substantially between 1969 and 1977.[16]

In fact, rank-and-file workers began to consider workplace safety and health a priority. Data are not available to assess changes over time on worker tradeoffs between safety and wages, or on attitudes about worker input into decisions over working conditions, but a third of the production workers surveyed in the 1977 QES indicated that they were willing to forgo a 10% increase in pay for "a little more" workplace safety and health. Most striking, over three-quarters believed that they should have "a lot" or "complete" control over safety and health. This was more than twice the percentage of workers who thought they should have this kind of power over any other issue, including how work was done, wage levels, hiring and firing, and hours and days of work. Significantly, rank-and-file interest in health and safety was *uniformly* strong among production workers.[17]

Union Activities

Several unions also made occupational safety and health an issue in collective bargaining and in politics. Union contracts in the auto, steel, rubber, and chemical industries reflected a new awareness; contracts signed during or in the immediate aftermath of the struggle over the OSH Act were especially innovative. The United Rubber Workers' 1970 contract with the major rubber companies established a five-year, company-financed, labor–management-run, university-administered research program in occupational health hazards in the industry. The UAW signed the first union contract with a national safety and health clause, and it closely followed the terms of the OSH Act. The contract specified worker rights to information about working conditions, including local union rights to receive industrial hygiene data and to take air samples; worker rights to receive reports on accident investigations; and worker rights to be given access to information on chemicals used in the plants. The contract also obliged employers to train union safety representatives at company expense. In 1971 the USWA negotiated company-paid joint committees and union inspection rights, expedited procedures for handling grievances of imminent-danger situations, and management recognition of its responsibility to respect the worker rights provisions of the act. Local 1557 of the Steelworkers won a detailed agreement designed to reduce worker exposure to coke-oven emissions at the Clairton Works in Pennsylvania, the largest coke-oven plant in the United States.[18]

In 1973 OCAW became the first union on record to bargain to an impasse over occupational health and safety. The union had attempted to win major concessions from the petroleum industry in 1970–1971, but had failed. In the next round of talks, OCAW refused to back down and ultimately forced the companies to agree to periodic company-financed surveys for work hazards by independent industrial health consultants selected with the approval of the union. Unlike the Rubber Workers, OCAW failed to win a company-financed health research fund. In other respects, however, the agreement was stronger than the URW contract because it specified a procedure for remedying hazards reported by the survey. A joint labor–management health and safety committee was established to decide on the implementation of corrective measures. If labor and management failed to agree on a plan, the union could take the issue to grievance or third-party arbitration.[19]

Many unions established or increased in-house health and safety activities. Before the passage of the OSH Act, only a few had active health and safety departments. By the early 1980s, 50 unions had ongoing efforts. The UAW, URW, Amalgamated Clothing and Textile Workers, International Union of Electrical Workers, United Electrical Workers, and International Chemical Workers joined OCAW, UAW, and USWA in creating local safety and health representatives. The UAW, OCAW, the Machinists, and the UMW developed training programs for local leaders. The UAW established an elaborate health and safety committee network, linking local efforts to the international's social security and health and safety departments. The Mineworkers established mandatory health education classes for approximately 3000 local committee members. The USWA and OCAW set up programs to monitor workplace hazards and published and distributed health and safety materials to their members.[20]

Some unions increased their staff and financial commitments to worker health and safety. The HRG conducted surveys of the health and safety activities of more than a dozen unions in high-health-hazard industries in 1976 and 1983 (see Table 5.3). During those eight years, the professional health and safety staff of these unions doubled; the ratio of full- to part-time staff increased by 28%. Health and safety budgets also grew. In 1976, 6 of the 15 unions surveyed spent less than $50,000 a year on health and safety; the minimum union expenditure rose to over $100,000 per year by 1983.[21]

Table 5.3.   Union Health and Safety Staff, 1976 and
1983, in Selected High-Hazard Unions[a]

Category	1976	1983
Doctors	5	3
Industrial hygienists	8	15
Lawyers	5	8
Engineers	5	4
Epidemiologists	1	2
Public health professionals	5	3
Economists	4	2
OSH specialists	0	4
OSH directors and coordinators	0	16
Nurses	0	1
Legislative representatives	0	1
Educators	0	1
Others	0	10
Chemists	2	0
TOTAL	35	70
FULL-TIME	20	44

[a]Staff is restricted to professional, technical, and supervisory
staff. In 1983 the 14 unions reported a category of additional
"union staff" that had no analogue in the original survey. Such
staff is generally composed of union staff members who have
received training in occupational safety and health issues but
are not specialists in the area. They have been omitted from
the table for the sake of cross-year comparison.
*Sources: 1983 Survey of Fourteen Union Occupational Safety
and Health Programs* (Washington, D.C.: Public Citizen Health
Research Group, 1984) for 1983 figures; *Survey of Occupational
Health Efforts of Fifteen Major Labor Unions* (Washington, D.C.:
Public Citizen Health Research Group, 1976) for the 1976 data.
Note that the sample changed slightly from 1976 to 1983. Only
14 of the original 15 unions responded.

The Labor–Environment Coalition

The unions' political efforts also showed signs of change. Most important, the links forged by labor activists with middle-class health and safety reformers in the 1960s remained and, in some cases, deepened. A coalition of environmentalists, public-interest groups, and labor representatives formed to lobby in defense of the social regulatory agencies, including OSHA, EPA, and CPSC. Eighteen national environmental, labor, and urban-action groups, including the Sierra Club, the UAW, the National Welfare Rights Organization, and the USWA, formed the Urban Environment Conference (UEC).

Reflecting an increased awareness of the need to overcome potentially divisive issues, unionists and environmentalists cosponsored a number of committees and organizations to promote the idea that environmentalism and jobs were not mutually exclusive. These efforts were critical to both movements. The postwar accord rested on a consensus on the necessity for economic growth. If industry or conservative trade unionists could convince workers that their jobs were threatened by health and safety regulations, efforts to forge new alliances, or even to implement existing protective legislation, were likely to fail. President Abel of USWA clearly recognized the problem and urged his colleagues in organized labor in 1971 not to succumb to "environmental blackmail." In keeping with this strategy, the Steelworkers supported several air-pollution regulations opposed by the steel industry.[22]

Most environmental groups and some unionists agreed, and the links between organized labor and the environmental movement were strengthened. Several new groups and coalitions were formed, including Environmentalists for Full Employment (EFFE), the National Committee for Full Employment (NCFE), and the Labor Committee for Safe Energy and Full Employment, organized by EFFE and several union staffers. The EFFE was founded by community and environmental activists "to publicize the fact that it is possible simultaneously to create jobs, conserve energy and natural resources and protect the environment." Cosponsoring groups called for a "general re-evaluation of our human and natural resource policies" and an end to the "exploitation that environmentalists and labor unions have, heretofore, fought independently." In 1976, with the UAW, UEC, and

National People's Action, EFFE held a conference on "Working for Environmental and Economic Justice and Jobs" at the UAW's Black Lake Conference Center. Attended by members from OCAW, the Sierra Club, the Natural Resources Defense Council (NRDC), and the Environmental Defense Fund, it was designed to develop the union–environmental network and educate activists in both movements about issues of common concern. Organized at the end of the decade, the Labor Committee for Safe Energy and Full Employment held conferences and published newsletters with the support of approximately a dozen unions, including the UAW, Machinists, and Mineworkers. Twenty trade unionists, including William Winpisinger, president of the Machinists, signed a letter against nuclear "blackmail" in 1979. In 1981, OSHA/Environmental Network was formed by national labor and environmental groups to monitor Reagan administration efforts to deregulate health and safety.[23]

More focused efforts also were tried. The Nader-sponsored Health Research Group (HRG) worked with unions to petition and sue OSHA to set and enforce standards. Led by Dr. Sidney Wolfe, the HRG was particularly active in efforts to force the agency to issue standards covering sanitary conditions for farmworkers, exposure to pesticides in the fields, cotton dust, and worker rights to information about hazards. The Women's Occupational Health Resource Center was established by Jeanne Stellman at Columbia University's School of Public Health. The Coalition for Reproductive Rights of Workers was formed to deal with workplace hazards that threatened fertility. The White Lung Association was formed to lobby and publish materials on asbestos.

Grass-Roots Struggles

There were also significant signs of radical change at the grass roots. Committees, Councils, Coalitions (and, in a few cases, Projects) on Occupational Safety and Health (uniformly referred to as COSHs) formed in the early 1970s to educate workers about health hazards and forge links among the rank and file, health professionals, and labor activists interested in issues of workplace democracy and worker participation. Begun by Chicago-area medical doctors in

1972, the COSH movement spread throughout the industrial areas of the Midwest and northeast. In 1980, at the height of the movement, there were approximately 20 COSHs, with especially active groups in Philadelphia (PHILAPOSH) and New York (NYCOSH). Most included union locals, individual workers, and health and labor activists.[24]

The COSHs were organized locally and remained independent organizations, coordinated informally by a network of health and safety activists. As a rule, they emphasized worker education and political action, although the specific mix of activities, as well as the emphasis given to national and local issues, varied by group. To educate workers, COSHs organized conferences, published fact sheets, and held seminars on the hazards faced by local unions. They also lobbied state and local governments for improved occupational health statutes, enforcement of existing laws, and reform of the workers' compensation system. Some COSHs focused on OSHA, serving as the agency's grass-roots watchdog. In 1980 the movement coordinated its efforts to help defeat S. 2153, a sweeping congressional amendment to the OSH Act that would have exempted approximately 90% of workplaces from OSHA inspections.

Many COSH members were New Leftists, and they brought a more radical vision to the problem of working conditions than most union officials did. The COSHs argued for worker control over working conditions and encouraged rank-and-file organization and participation as they lobbied for legislative reforms and educated union locals about health hazards. After 1977, under Dr. Eula Bingham's leadership, OSHA provided financial support to the COSH movement through the New Directions Grant Program. This program was designed to fund private efforts by business, labor, and nonprofit organizations to "increase employer and employee awareness of occupational safety and health." The program was small; its yearly budgets were in the $3 million to $4 million range and spread over a large number of organizations. But nearly half of the COSHs received funding, including groups in Chicago, Philadelphia, Maryland, Massachusetts, North Carolina, New Jersey, Rhode Island, and New York.[25]

Significantly, grass-roots efforts and innovative forms of direct action by rank-and-file workers, local unionists, and community activists often produced results that OSHA could not. The Philadelphia COSH,

PHILAPOSH, compiled a list of examples taken from the Pennsylvania area that is worth considering at some length.[26]

■ United Electrical Workers Local 141 conducted a mass-education campaign among workers at Cooper Industries' Penn Pump plant in Easton, Pennsylvania, that led the company to change the solvent used in production and improve the spray booths where workers applied it.

■ The chief steward of Local 111 of the International Union of Electricians took worker complaints about exposure to PCB to a local television station. In response, the company industrial hygienist ordered new handling procedures and started medical exams for employees already exposed to the substance.

■ AFSCME Local 2187 in the Philadelphia City Health Clinic forced the city to remove asbestos from ceilings after local union officers threatened to leaflet patients in the clinic about the danger.

■ Pennsylvania Service Employees Local 668 of state welfare workers used mass picketing to force management to reduce workloads after employees complained of chronic stress.

■ UAW Local 1612 at the Gould assembly plant in Philadelphia held a "nurse-out" to force the company to stop using a new fiberglass-coated wire suspected of causing itching and skin rashes. With 50 to 60 workers per day leaving their posts to be examined by the nurse, the company replaced the wire.

■ An OCAW local in a Tenneco plant in New Jersey used a "lunch-out" (they refused to eat in the company lunchroom and ate on the lawn instead) to protest the company's failure to correct illegal levels of lead, cadmium, and carbon black found by OSHA. After local media and community residents became involved, the company made a variety of concessions, including new contract language on informational rights, health and safety training, and a new grievance procedure in which the burden of proof was shifted to the company to prove that an alleged hazard was safe.

■ A United Electrical Workers local at Dynamic Products in Middletown, Pennsylvania, protested cold conditions in the plant by working only in warm areas. When the company refused to pay the workers, they successfully claimed unemployment insurance. Management installed new heaters.

■ An International Chemical Workers Union local in Boyertown, Pennsylvania, adopted a policy of making all complaints about working conditions in writing. In 1980 they filed 697 complaints with the company and threatened to take each through the grievance procedure or to OSHA.

■ UAW Local 1612 filed 35 grievances based on existing OSHA standards and overwhelmed an unprepared management, which then negotiated with the union.

■ United Electrical Workers Local 168 tested for and found dangerously high vapor levels and forced the Rois Manufacturing Company to install a fan and hood.

■ Seven hundred employees in a Communications Workers local in Camden County, New Jersey, protested the lack of ventilation in the summer by walking out and urging reporters to take temperature readings. The company fixed the ventilation system and paid the workers for the time they were off.

■ USWA Local 4588 at Budd's trailer plant in Eagle, Pennsylvania, negotiated the right of the safety committee to "red tag" machines that cause imminent dangers or to call in an international union representative to do the same.

Similar examples can be cited from other states and unions. But in almost all cases the pattern is the same: workers at the local level were able to use sympathetic media and health and safety activists to force firms to comply with OSHA rules or make changes not covered by OSHA regulations.[27]

The Reproduction of Economism

For the most part, however, the possibility of radical action on this issue went unrealized. Despite the impact of grass-roots efforts and the good intentions of some union leaders, the labor movement did not take full advantage of the opportunities afforded it by health and safety reform. Instead, it retained its economistic strategy and focused its money and time on defending the organizational interests of the unions and the economic interests of the rank and file. Control over work and the labor process was ceded to employers.

As a result, many of the more promising changes that initially occurred in union lobbying and collective bargaining were not sustained, and links to other movements remained superficial. Most unions continued, or returned to, traditional bargaining and political strategies. They defended OSHA, but their defense was unimaginative, uninspired, poorly funded, and often ineffectual.

To some extent, the decision to pursue conventional strategies was a nondecision; unions continued to do what they had done for so long. But several of the more health- and safety-conscious unions grappled with the strategic issues raised by regulation and the business attack on it, and a few recognized at least some of the stakes involved. From the beginning, USWA leaders anticipated that social regulation's opponents would attempt to drive a wedge between environmentalists and labor. They also understood the importance of involving members in the defense of health and safety. The USWA's in-house analysis of the success of the anti-OSHA forces concluded that "the lack of grass roots political support . . . indicates an area in which our own approach must change. *The workplace environment must become a politically emotional issue among our members*" (emphasis in original).[28] The UAW recommended a four-point response to attacks on OSHA, including strengthening the local union programs, involving members in in-plant programs, and developing "grassroots coalitions with environmental and consumer groups, activist medical and public health students and faculty, concerned professionals and other labor unions at the local level."[29]

Even though these recommendations were modest, most unions found them to be more than they could or wished to follow. This is evident on a variety of fronts. Collective bargaining provisions covering health and safety, for example, remained relatively constant before and after the OSH Act. The 1975 study for the Bureau of Labor Statistics cited earlier indicates that, at best, modest overall gains were made in specific rights to safety and health after 1970. Equally important, many unions had failed to learn the lessons of the OSH Act. Few of the existing provisions unambiguously granted worker rights, clearly detailed company responsibilities, or empowered joint labor–management health and safety committees to establish and imple-

Table 5.4.   Changes in Collective Bargaining Before and After 1971

Provision	Percentage of agreements on references to health and safety that contain selected provision, rank ordered by percentage change		
	Before 1971	**After 1971**	**Change**
Discipline for noncompliance	18.5	25.5	37.8
Employer pledges of compliance with law	41.4	53.2	28.5
Safety inspections	13.2	15.3	15.9
Employee rights with regard to safety	23.5	26.6	13.2
General policy	39.8	45.0	13.1
Joint safety committees	19.5	21.5	10.3
Safety committees	22.9	24.7	7.9
Union-management cooperation pledges	20.6	22.0	6.8
Safety equipment	55.0	57.9	5.3
Union rights with regard to safety	15.8	16.3	3.2
Accident procedures or compensation	63.0	64.5	2.4
Physical examinations	21.8	21.3	-2.3
Sanitation provisions	51.9	50.0	-3.7

Source: Adapted from Winston Tillery, "Safety and Health Provisions Before and After OSHA." *Monthly Labor Review* 98, no. 9 (1975), table 2, p. 42.

ment safety rules and work practices independent of approval by plant supervisors and higher authorities. Employee rights to physical examinations actually declined slightly (see Table 5.4).

Union staff and budgetary increases were also less significant than they appear at first sight. Base-year totals were small and exaggerate the rate of increase. Most of the new staff was concentrated in four unions: the mineworkers, OCAW, and the painters' and paperworkers' unions. Only five unions reported more than one staff person per 50,000 workers; only six reported spending more than a dollar per worker per year on health and safety. As the HRG concluded, these unions "still employ a largely inadequate number of health and safety personnel," and few showed "significant staff improvements over the past seven years."[30]

Most telling, the unions continued to use their political resources to lobby as they had done throughout the postwar period—as interest groups and clients of the state rather than as advocates of radical changes in the organization of production or the relationship between the state and economy. The labor movement did not question the liberal approach to work that it, and the act, had taken. To the AFL-CIO and the individual unions that pressured the agency and Congress, the issue was one of implementation. They demanded full enforcement of the act and individual standards for particular groups of workers. Specific issues varied over time: in the early 1970s the unions vehemently protested the Nixon administration's effort to allow the states to take back enforcement of the act without adequate supervision; in the mid 1970s they concentrated their efforts on blocking amendments to the OSH Act that would have exempted small business from enforcement; in the late 1970s they lobbied for higher penalties, stricter standards, and funding for union education programs. As a rule, however, the labor movement restated a traditional theme: the agency's failures resulted from limited resources, poorly designed programs, and bad faith on the part of political leaders.[31]

In most cases involving standards, the unions engaged in what can be called "hazard-driven" lobbying: individual unions or ad hoc coalitions of unions demanded and defended policies that protected *their* workers. The Textile Workers lobbied for the cotton-dust standard; the Rubber Workers lobbied on benzene and acrylonitrile; the Steelworkers lobbied on lead, coke-oven emissions, and chromium; the UAW lobbied on noise. Multiunion coalitions formed only when hazards affected a cross-section of workers, as was true with the noise standard when the UAW, IAM, and USWA worked together to lobby the agency.

Although strategically placed to do so, the AFL-CIO failed to exercise significant leadership in these efforts. To the contrary, reprising its efforts of the 1960s, it played a marginal role, content to take credit but reluctant to contribute. The federation mounted occasional "OSHA Watch" mobilizations, but these were perfunctory. Its health and safety staff remained small—one full-time industrial hygienist and one full-time director through 1983; one full-time director after 1983. It monitored rather than directed union activity and took its lead from more aggressive unions, principally USWA and UAW, and public-interest lobbies, notably HRG.

While broad multi-interest coalitions formed to defend the OSH Act and other health and safety statutes, these efforts did not result in the creation of permanent political organizations that institutionalized the health and safety movement or linked it to a wider political movement for reform of the economy. Environmentalists' efforts to forge a coalition with labor in the mid 1970s around jobs and environmentalism were frustrated by labor's economism. Pronuclear, the building and construction trades were particularly hostile to the environmental movement and its concern for safe and renewable energy sources. The AFL–CIO, dominated by these unions, followed suit and rejected the movement's criticism of organized labor's growth-at-any-price philosophy. As George Meany's aide reportedly told Black Lake conference organizers, "Where exactly the energy comes from for [these jobs] is not a big issue with trade unions."[32] Efforts to link the issues of nuclear energy and nuclear arms, or even to focus attention on the hazards of radiation to workers, ran into the same kind of cold-warrior opposition that frustrated attempts to publicize the hazards of uranium mining in the 1960s.

Although the UAW and OCAW continued to support the EFFE after Black Lake, even the liberal unions failed to do all that they could to promote stronger ties. Apparently, the Old Left in the labor movement saw these issues as marginal or frivolous. The Progressive Alliance, a UAW project designed to provide the more liberal unions with an independent political voice, excluded all but one environmental group from its board of directors.[33]

Broad coalitions also formed in response to economic review during the Carter administration, but they were ad hoc and did not change the way that groups seeking protection approached the state. These groups typically included the OCAW, USWA, Sierra Club, National Wildlife Federation, NRDC, EDF, Consumers Union, and National Consumers League. No permanent umbrella organization emerged to coordinate the efforts. Indeed, several of the largest unions moved in and out of these coalitions as the particular issue changed. Labor leaders contributed their names and organizational affiliations, but little more.

Having failed to link up with other movements, or to take its own radical rank-and-file movements seriously, organized labor was unable to offer a positive vision of worker health and safety that tied the gains made in the OSH Act to a new vision of how work might be organized or the economy restructured. As a result, although they defended social regulation, neither the federation nor individual unions were able to do what business had done—link their interests to a compelling societal interest. Instead, organized labor simply asserted that workers had rights to protection and that these rights were natural and incontrovertible. Or, in the words of the AFL–CIO's legislative director, health and safety was "as basic a right as any of our freedoms." [34]

Thus the labor movement was unable to respond imaginatively to the claim by business that resource constraints had to be taken into account when selecting hazards for control. The OSH Act's critics were able to portray demands for worker health and safety as another costly, "special" interest, paid for by other Americans. Public opinion was sympathetic to occupational safety and health regulation, but without a positive vision that linked self-determination at work to democratic control over the organization of production generally, unionists found themselves on the defensive, and ineffectual in the face of organized opposition to reform.

Ultimately, the unions' conservatism also helped undermine the COSH movement; despite their innovative approach to health and safety, the COSHs withered for lack of union support. True to their radical origins, most COSHs found it difficult to work easily with union leaders. For their part, many national unions feared the impact of the COSH movement on their organizations. Despite its strong commitment to OSHA, for example, the Steelworkers initially opposed the establishment of the Pittsburgh COSH because it feared its impact on employee–union relations. Many USWA locals eventually became actively involved in the Chicago-area COSH, but national unions remained suspicious of the movement. Some COSHs excluded unions. For example, NYCOSH did not admit union locals as members until the end of the 1970s, and its membership consisted of individual health professionals and union staff members.

Consequently, the COSHs developed alongside rather than within organized labor. By the mid 1970s, most COSHs recognized the costs of political isolation in an increasingly hostile political environment. In response, many forged closer working relationships with union leaders. But to gain national union support, the COSHs moderated their demands and narrowed their activities to conform to union approaches to work. Only a dozen or so COSHs survived intact into the 1980s, and these were pale reflections of the original radical vision. They continued to do important work, but the last bridge between conventional labor liberalism and the more radical vision of the original health and safety reformers had been seriously weakened.[35]

Why Organized Labor Failed to Take the Initiative

Several things help explain the unions' failure to alter their approach to work and the economy. The structural organization of the political economy remained in place and undoubtedly helped to shape worker demands. As I noted earlier, workers' economic dependence on private investors and employers leads them to moderate their demands. This tends to discourage movements for health and safety and encourage unions to focus their attention on traditional economic issues. Clearly these incentives remained in place. Worker attitudes did change, however, and there were signs of more radical grass-roots activities. Three additional factors account for the unions' failure to marshall this discontent and use it to strengthen their position.

Perhaps most important, the labor movement entered a period of decline in the 1970s. Union membership fell precipitously from the late 1960s to the early 1980s—from 25% of the labor force in 1969 to 18% in 1982. The number of workers in unions actually began to decline absolutely in 1978; by 1982, there were fewer workers in unions than there had been in 1969. As membership declined, unions found it difficult to win at the bargaining table. Although powerful unions such as the UAW and USWA were able to hold their own throughout the 1970s, other unions found it harder and harder to negotiate significant wage increases or, more important, to expand the scope of bargaining rights. By the 1980s, most unions, including the UAW and USWA, were making contract concessions on wages and work rules. The unions in manufacturing industries, such as the UAW, USWA, ACTWU, and URW—the unions most active on occupational safety and health issues—were particularly hard hit.[36]

The union decline was also reflected in the waning of rank-and-file militance, union strike activity, and union political power. Wildcat strikes, for example, went from a peak of 39.8% of all strikes in 1972 to 13.4% in 1980. The number of strikes called to win improvements in wages and benefits fell, while the proportion of "defensive" strikes—called to protect existing rights and benefits—rose.[37] The unions' clout on Capitol Hill followed a similar trajectory. From the mid 1970s onward, the percentage of members of Congress supported by COPE or with voting records favorable to organized labor fell from nearly two-thirds in both the House and Senate to less than a majority by 1980.[38] Nowhere is the unions' declining political power more evident than in their unsuccessful fight in 1977 and 1978 to pass a labor law reform bill—organized labor's top legislative priority—despite nominal support from the Democratic party and President Carter.

The causes of the union decline are complex, but five factors stand out. First, the cumulative effect of a decade of high unemployment undoubtedly weakened the labor movement. Second, competition from international sources and from nonunion domestic firms also undercut workers' bargaining position. Third, government policy reinforced these factors, particularly after 1981 as the Reagan administration embarked on a campaign to reduce social spending and discourage labor militance. Cuts in unemployment insurance payments, for example, increased the economic costs to workers of job loss, while the government's hard-line position in the airline controller (PATCO) strike and antiunion appointments to the NLRB raised the risks of job loss for striking workers. Fourth, many corporations and industries shifted from a policy of accommodation with labor to a campaign to roll back unionism. Just as employers efforts against OSHA became more class conscious and coordinated, antiunion efforts became more militant and effective.

Finally, the labor movement's political and economic strategies failed to confront the new situation. Instead of organizing nonunion workers or reaching out to new constituencies, most unions sought

to defend their own organizational and economic interests and eschewed broad-based economic and political mobilization. In regard
to occupational safety and health, this meant politics as usual, as I
outlined above. The labor movement lobbied Congress and the White
House to increase OSHA budgets and pressured OSHA officials to
enforce the act. It policed the agency rather than mobilized workers.

The consequences of the union decline for occupational safety and
health were immediate. As membership dropped and unions failed
to win wage increases, their resource base declined. All union activities suffered, particularly those, such as health and safety, that were
traditionally considered secondary issues. Many unions refocused
their efforts on fighting battles that they thought had been won, including defending themselves against business attempts to win de-
certification elections. Union attention to health and safety in collective bargaining peaked between the passage of the OSH Act and
1975; subsequently, most unions devoted more attention to jobs and
union security and less to the "quality of life" at work or in the environment. The AFL-CIO maintained its OSHA Watch and cooperated
with environmentalists and public-interest groups in coalitional work,
but little attention and even fewer resources were devoted to these
efforts.

An Ideological Defeat

By the late 1970s, business had retaken the ideological offensive and
organized labor had failed to resist the assault. Labor's conventional
strategy proved less and less successful as economic conditions
worsened and business pressures mounted. As a result, social rights
had less resonance on Capitol Hill and in the White House. Business
lobbyists were not able to win everything they wanted in the 1970s. As
we see later, Carter was only partially sympathetic to their demands;
the unions were able to keep OSHA alive and growing, albeit slowly,
until the end of the decade. But among elites and opinion leaders, the
climate had shifted against organized labor and workplace reform.
In the next chapter we look at how this shift resonated through the
White House and the courts.

[6]
The White House
Review Programs

he OSH Act's right to protection
rests on OSHA's ability to design
and implement policies that give
it force; by placing OSHA in an
executive department, however,
the law made it likely that business would lobby the White House
to curtail occupational safety and
health regulation. As Chapter 4
described, business opposition to
social regulation in general, and
the OSH Act in particular, was intense as business
groups mobilized in new ways to block the implementation of
these reforms.

Predictably, employers took their case to the executive
branch, and as the balance of political forces shifted against
the idea of worker protection in the 1970s, the White House
responded. Undoubtedly the well-funded and well-organized
business campaign helped shape presidential policy toward health
and safety regulation. But the White House had its own reasons for
seeking to control OSHA. As I discussed in the Introduction, the capitalist organization of production generally constrains public policy;

economic crisis in the 1970s heightened this constraint. After 1973, as the economy weakened, Presidents Ford, Carter, and Reagan used their executive powers in an effort to force OSHA to take economic factors into account.

Designed to assess and minimize the economic impact of social regulation, the resulting White House review programs involved unprecedented presidential supervision of administrative regulation. Before the 1970s, presidents were reluctant to become directly involved in regulatory agency decisions. Most agencies are distant from the Oval Office and protected by political and legal norms of deference to administrative expertise and nonpartisanship. Most agencies are also defended by well-organized constituents who will mobilize to prevent unwanted interference with rulemaking. Under these circumstances, the benefits of intervention are not likely to be large, and the potential costs are high. As a result, agencies are generally left to negotiate policy among affected groups in their immediate environment, and White House officials, particularly the president, let them alone. But as this chapter chronicles, after 1973, business mobilization and economic crisis led to a fundamental change in the relationship between the president and the social regulatory agencies, and severely limited OSHA's policy options.

Economic Crisis and the Problem of Business Confidence

The concentration of control over productive resources in the hands of private investors and employers constrains public officials who seek to regulate them. This constraint is probably most effective when economic conditions are uncertain and profit rates are low or declining. Then, elected leaders are apt to solicit business confidence by pursuing policies that promote corporate profitability.

One way to do this is to reject reforms that challenge business interests. Or, if these reforms are already law, they can be "rationalized" administratively. That is, program implementation can be crafted to take costs and managerial prerogatives into account—to respect the needs and logic of private capitalist investment. The trajectory of White House intervention into social regulation after 1973 suggests

Figure 6.1. Rate of Unemployment, Civilian Workers, 1965–1984

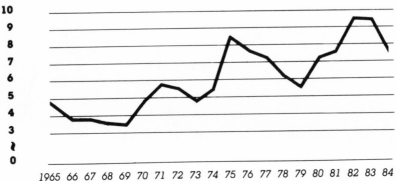

Source: *Economic Report of the President, 1985* (Washington, D.C.: GPO, 1985), table
B-29.

that this did in fact occur. A decade of economic recession and sluggish growth led successive administrations to develop an economic
review program that they then used to protect business interests from
health, safety, and environmental regulation.

That the review program developed at least in part in response to
structural forces is suggested by the close correspondence between
changes in the state of the economy from 1971 to 1984 and changes in
White House policy toward social regulation. The economic crisis of
the 1970s and early 1980s is familiar: after 1973, economic decline
replaced the high-spirited growth of the mid and late 1960s, and
despite several cyclical recoveries, sluggish growth remained a fact
of economic life through 1982.

Figures 6.1, 6.2, and 6.3 indicate the scope and severity of economic
problems in this period by displaying the rates of unemployment,
inflation, and capital investment between 1965 and 1984. The rates of
unemployment and inflation indicate that private investors had good
reason to be wary of the American economy in the 1970s, and investment rates suggest that they were. The economy was deeply
disturbed.

Figure 6.2.   Rate of Inflation, 1965–1984

Inflation rate

Note: Rate is the year-to-year change in the consumer price index.
Source: *Economic Report of the President, 1985* (Washington, D.C.: GPO, 1985), table B-56.

White House Oversight of Social Regulation

As a rule, the president's economic advisers watch trends closely and recommend policies to counteract declining business confidence. Normally, they concentrate on conventional economic levers, such as budget deficits, tax and interest rates, and the money supply. The timing of the White House review programs, however, suggests that their concerns were felt in this area as well.

The executive branch took the first step toward presidential review of agency rulemaking during the first half of the Nixon administration as inflation surged to new heights, but the program did not sink institutional roots until the 1973–1975 recession. In 1974 Nixon singled out regulation in his last public speech on inflation and ordered the Office of Management and Budget (OMB) to make a "sweeping

Figure 6.3.　Investment in the U.S. Economy, 1965–1984

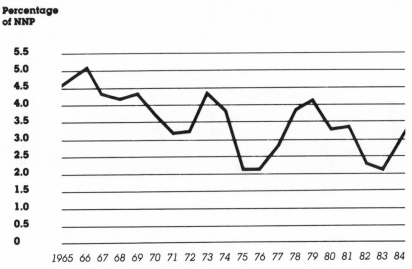

**Percentage
of NNP**

Note: Percentage of net national product (NNP).
Source: Economic Report of the President, 1985 (Washington, D.C.: GPO, 1985), table B-29.

review" of regulations that restricted supplies and fed inflation.[1] On taking office after Nixon's resignation, Gerald Ford issued the first Executive Order requiring agencies to submit Inflation Impact Statements (IISs) to OMB, created the Domestic Council Review Group on Regulatory Reform to coordinate White House efforts, and called the first Regulatory Summit Meeting of agency heads and White House policy advisers.

From 1974 through 1983, each administration built on its predecessor's efforts to augment the White House's oversight powers. Control over social regulation was centralized in the Executive Office of the President (EOP), and the economic evaluation of health, safety, and environmental standard-setting and enforcement activities was systematized. Carter followed Ford's Executive Order with one that required Economic Impact Statements (EISs) and created two additional White House agencies: the Regulatory Analysis Review Group (RARG) and the Regulatory Council (RC), to administer the program.

Table 6.1.   Major Regulatory Review Events

1971   Quality of Life Review Program initiated.

1974   Ford E.O. 11821 requires Inflation Impact Statements.

1975   Domestic Council Review Group on Regulatory Reform
established.
Regulatory Summit at White House.

1978   Carter E.O. 12044 orders regulatory analyses.
Regulatory Analysis Review Group established.
Regulatory Council established.

1981   Reagan's 60-day freeze on all pending regulations.
Reagan E.O. 12291 requires Regulatory Impact Analysis.
Office of Information and Regulatory Affairs created.
White House Task Force on Regulatory Relief established.

1983   Task Force on Regulatory Relief abolished.

Reagan strengthened the program by requiring Regulatory Analyses (RAs), creating the Office of Information and Regulatory Affairs (OIRA) in OMB to coordinate all oversight programs, and establishing the White House Task Force on Regulatory Relief to supervise the entire deregulation campaign (see Table 6.1).

Even the cyclical movements of the economy correspond to variations in the timing and intensity of the White House review effort. The fairly strong economic recovery between 1976 and 1979 is reflected in the failure of either Ford or Carter to make social regulation an issue in the 1976 presidential election. That the economy expanded at a healthy rate during the first year and a half of Carter's term helps explain why he felt free to respond to labor pressure and moderate the review program in 1977. Carter's increased interest in cost control after 1978 corresponds to a rising rate of inflation in the second half of his administration. Even the development of the Reagan review program corresponds to cyclical economic changes. Along with tax relief, the administration offered deregulation as one of its principal economic recovery measures in 1981 and pressed the program for two years. Murray Weidenbaum was appointed chair of the CEA; OMB reviewed thousands of regulatory proposals; hundreds of major actions were delayed or discarded. But in 1983, with recovery at hand, the Task Force on Regulatory Relief was abolished, and deregulation was dropped from Reagan's agenda.

Of course, care should be taken in analyzing these short-term correlations. White House aides were at work on policy proposals long before they became public and programs were implemented. Short-term political forces undoubtedly affected the timing of various initiatives as well as the intensity of White House support. Reagan's general ideological opposition to government regulation of business certainly played a greater role in his support for economic review of social regulation than short-term changes in economic activity. The success of his deregulation campaign probably eliminated the need to strengthen the program after 1983. Nonetheless, the relationship between the development of the White House review program and changes in the economy suggests a clear relationship between presidential concern for business confidence and the subordination of social regulation to White House review.

Capitalist Symbols

An analysis of the ways in which the economy constrains public officials argues that efforts to rationalize social policy be viewed as a form of symbolic politics. Two facts indicate that the White House's review program was a symbolic concession designed to restore business confidence in troubled economic times.

First, social regulation does not seem actually to account for a significant part of the post-1973 economic decline. The Weidenbaum study showed that regulatory compliance accounted for 4% of the GNP and more than one-third of all private investment in new plant and equipment.[2] From these figures, Weidenbaum concluded that the macroeconomic effects were likely to be debilitating. More careful studies of the impact of regulation on productivity, economic growth, employment, and inflation have reached more modest conclusions. Edward Denison's work on the slowdown in productivity from the late 1960s through the mid 1970s concluded that changes in the "legal and human environment" between 1973 and 1976 accounted for only 12.5% of the 3.2% decline in the growth in national income per person during that period. Pollution-control expenditures accounted for more than half of this small share; worker health and safety accounted for approximately 30%.[3] Other studies, using slightly different methodologies, have come up with slightly different estimates, but none supports Weidenbaum's belief that regulation was strangling free enterprise. William Nordhaus, summarizing this literature, offers a "best guess" estimate that it accounted for approximately 10% of the productivity decline.[4]

Other studies suggest that even these estimates are overstated.[5] Even if we accept Denison's figures, they do not show that deregulation would do much, on its own, to increase productivity or reduce inflation. More important for our purposes, OSHA itself played a very small part in the overall picture. Since its regulations accounted for less than one-third of the productivity gap that Denison attributed to regulation, it caused, at most, 3.8% of the total productivity shortfall.

As for capital diverted from more "productive" uses, OSHA regulations absorbed only 2.1% of total capital spending per year in the manufacturing sector between 1972 and 1983 (see Figure 6.4). To be sure, some standards were costly. The textile, electrical instruments, transportation, rubber, iron and steel, and chemical industries all increased capital investment in worker health and safety at some point after 1971 to come into compliance with agency standards. But only one of these industries devoted more than 10% of its total capital investment in any single year to OSHA-related activities. In general, capital spending on OSHA was minimal.[6]

Nonetheless, public officials understood the psychological importance of economic review and self-consciously used the program to communicate to business their commitment to economic recovery, sound public finance, and the verities of capitalist economics. As Carter's top economic advisers wrote him shortly after his inauguration, OSHA was important because it was the "leading national symbol of overregulation." "Not to act decisively," they warned, "would be perceived outside the labor movement as a retreat from your commitment to major regulatory reform." Changing OSHA was a way of showing nonlabor groups that "the composition of the reform effort reflects their concerns."[7] White House aide Simon Lazarus, in charge

Figure 6.4. Worker Health and Safety Investment as Percentage of Capital Investment in Manufacturing, 1972–1983

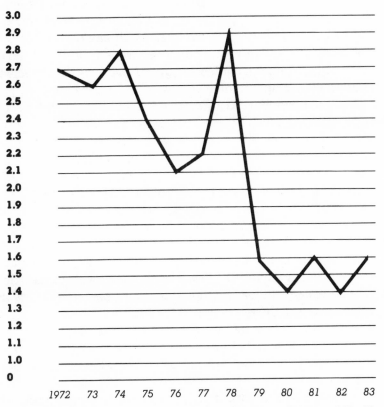

Percentage of capital spending

Source: *Annual McGraw-Hill Survey of Investment in Employee Safety and Health* (New York: McGraw-Hill, 1984).

of regulatory reform for Carter's Domestic Policy Staff, was especially frank: "No one is silly enough to pretend that these efforts will affect the consumer price index by half a point," he observed. But regulatory reform was a way of showing "that government is doing what it can to restrain inflation."[8] In a similar spirit, OMB Director David Stockman wrote Reagan that regulatory relief was important for the "long term signals it will provide to corporate investment planners."[9]

Deregulation as Economic Policy

To signal their concern about business confidence in the American economy, Presidents Ford, Carter, and Reagan made deregulation a major part of their economic policies. Although they often couched their criticisms in conciliatory tones, they paid close attention to what industry was saying, and White House rhetoric mirrored industry's concerns: protective regulation was blamed for economic decline, and society's general interest in capital investment was counterposed against the particular interests of workers in protection.

As Republicans, Ford and Reagan were more critical of social regulation than was Carter. They were quicker to blame regulation for economic decline and they were quicker to question the basic goals and methods of protective legislation. Ford's efforts were path-breaking in two ways. First, he focused the reviewers' attention on the impact of regulation on the private sector rather than on federal budget outlays, the traditional concern of "cost-conscious" Republican administrations. To this end, Ford issued an Executive Order that empowered OMB to review all regulations "which may have a significant impact on inflation." [10] Second, Ford argued explicitly that society's general interest in investment and growth justified deregulation. Social regulation was, he maintained, one of the prime determinants of almost every economic ill plaguing America, from inflation to unemployment to the decline of innovation, entrepreneurial enterprise, and individual liberty. Using makeshift, exaggerated cost estimates, he claimed that regulation cost Americans $2000 per family per year. [11]

The Reagan administration took up the cause and made deregulation—by then a household word—a centerpiece of its conservative revolution. During the presidential campaign, Reagan made his intentions toward OSHA clear in an interview with the *Washington Post*.

> My idea of an OSHA would be if government set up an agency
> that would do research and study how things could
> be improved, and industry could go to it and say, we have a
> problem here and seem to lose more people by accidents

in this particular function. Would you come look at our plant,
and then come back and give us a survey.[12]

After the election, he implemented his views of regulation in general
and OSHA in particular. Before a joint session of Congress, the President repeated the verities. America had experienced "a virtual explo-
sion in Government regulation" resulting in "higher prices, higher
unemployment, and lower productivity growth."[13] All pending regulations were "embargoed" and the Task Force on Regulatory Relief
was established.

Vice-President George Bush, the task force's chair, spelled out the
administration's goals in greater detail:

This administration considers regulatory relief one of its major
economic initiatives designed to restore the productive
capacity of the economy in order to create work for the jobless
and lower the rate of inflation that is devastating every
American, especially the poor and middle income workers.[14]

Regulatory relief, Bush maintained, could "expand the opportunities,
the productivity and the earning power of every American and every
segment of the economy."[15]

Carter was much less hostile to the principle of government regulation or the goals of the new agencies. He said little about the issue in
the 1976 campaign, and his transition team all but ignored it.[16] When
Carter addressed the problem of health, safety, and the environment,
he seemed sympathetic. He urged reforms that would make compliance programs more efficient and reduce the paperwork burden
on business; he recommended changes that streamlined enforcement
and relied on incentive systems to make compliance more cost
effective.[17]

As a result, although they were intent on regulatory reform, Carter's economic advisers avoided a frontal assault on the agencies.
They focused their criticisms on regulatory methods instead. Led by
Charles Schultze, an academic economist and CEA chair, they convinced Carter that what Schultze had named "command-and-control"
programs were obstacles to economic growth and entrepreneurial-

ism. In 1977 Carter formally endorsed this view. The OSH Act, he told a Yazoo City Town Hall meeting, was a "good piece of legislation." But its "emphasis on detailed regulations on safety" was "too much." [18]

The Consolidation of White House Control

Given the particular features of the American system of government, the effort to involve economic policy advisers in agency decision making required significant changes in the relationship between the White House and the agencies. Several affected agencies, notably the Consumer Product Safety Commission (CPSC) and the Federal Trade Commission (FTC), were independent commissions; White House leverage over them was limited. But even executive branch agencies such as OSHA had traditionally enjoyed a good deal of discretion. While the president was ultimately accountable for their decisions, the agencies' authorizing statutes gave the various departments direct responsibility for agency policy.

The OSH Act authorized the secretary of labor to set and enforce standards or delegate that responsibility to an assistant secretary. The president could assert control by hiring and firing the secretary or by supporting or rejecting agency budget requests. But the act did not give the White House direct control over OSHA decision making. Thus the development of the review program required that new, more centralized institutions be created to increase presidential supervision of day-to-day agency policy. The same held true for the National Highway Traffic Safety Administration in the Department of Transportation, or the EPA, an independent executive agency.

Just as White House efforts to control the budgetary process had led to the growth of the Bureau of the Budget a half century earlier, White House efforts to control regulation led to the expansion and reorganization of the Office of Management and Budget in the 1970s and 1980s. The first efforts to supervise the EPA were carried out by the "budget" side of OMB. Then, in 1974, regulatory oversight was transferred to OMB's "management" side; subsequently, regulatory oversight capacity increased dramatically. In 1981, OIRA was created to coordinate the entire oversight process. By 1984, OMB exercised nearly complete control over agency actions.

The precise role of OMB, including its relationships to other executive agencies interested in overseeing social regulation, varied by administration. Before 1981, OMB shared power with several other EOP organizations. These included the Council of Wage and Price Stability (CWPS), the CEA, and, during the Carter administration, RARG and the RC. After 1980, OIRA played the dominant role, supplemented by the White House Task Force on Regulatory Relief.

Overall, however, the institutional developments followed a fairly constant trajectory. White House oversight was increased and formalized through Executive Orders and supplementary OMB circulars that required executive agencies and departments to assess the economic consequences of proposed actions, keep OMB and other reviewing bodies abreast of proposals, and justify their proposals to OMB. In this way, the EOP gained greater and greater control over social regulation.

Ford's Executive Order required agencies to submit Inflationary Impact Statements, modeled on Environmental Impact Statements required by the National Environmental Protection Act, for all major regulatory actions. The OMB and CWPS monitored agency compliance with the Executive Order and filed formal written comments on agency inflation-impact statements. The latter, created and lodged in the EOP by Congress in 1974 in response to the president's request for increased authority to fight inflation, entered formal advisory statements in agency rulemaking proceedings.

Carter's Executive Order expanded the review process. Agencies were required to justify their actions with Economic Impact Statements that included "regulatory analyses" of the problem they proposed to tackle. These had to specify the alternative solutions considered by the agency; the economic consequences of each alternative, including their costs and benefits; and a comprehensive explanation of why the agency chose the alternative it proposed. Oversight was shared among a larger number of advisers and agencies, and the CEA played a greater role than it had in the previous administration. The RARG was created to represent the major executive branch agencies with "economic" responsibilities, including the DOL, Treasury, Commerce, OMB, and the CEA. The CWPS continued to file written comments under Carter; but RARG supplemented CWPS efforts by focusing the attention of EOP officials on a small number of particularly controversial regulations, thereby establishing a climate of heightened scrutiny.

After 1980, oversight was augmented, and OMB's powers were expanded. Both Ford's and Carter's efforts emphasized cost-effectiveness rather than cost-benefit tests, and in both administrations the reviewers' recommendations were advisory. Under Reagan, the emphasis of the review program shifted from regulatory "reform" to regulatory "relief" for business. Reagan replaced RARG with the White House Task Force on Regulatory Relief, similar in composition to RARG, but more visible and highly placed within the EOP.[19]

The Reagan review mechanism was much more elaborate and harder to satisfy than earlier programs. All executive agencies were required to justify rules with cost-benefit analyses. Agencies also were required to select regulatory objectives that "maximize the net benefits to society" and choose rules "involving the least net cost to society." Regulations not only had to pass cost-benefit tests but had to meet a restrictive definition of cost effectiveness imposed by the White House.

Reagan's Executive Order also formalized OMB's oversight powers. Agencies were required to secure OMB approval before publishing pending proposals and impact statements, renamed Regulatory Impact Analyses. The order also gave OMB the authority to specify precisely how these analyses were to be done. All these powers were lodged in OIRA, and OIRA's staff and resources were substantially increased, while OMB refused agency requests for additional funds to hire more impact-analysis personnel.[20]

Although it could not legally decide agency rules, OIRA's supervisory powers were so extensive that it wielded an effective veto over rulemaking. It could reject agency requests to publish proposals and final regulations until they satisfied its analysis requirements. It also had complete discretion to define those requirements. At the same time, OIRA could choose to waive these requirements if it so chose, making it possible for the administration to "fast track" deregulatory efforts that could not meet cost-benefit tests. Thus, supplemented by the 1980 Paperwork Reduction Act, which required OMB approval for all agency actions that imposed significant information-reporting requirements on affected organizations, the Executive Order made the White House the final arbiter of agency actions, qualified only by those statutory obstacles in the agencies' authorizing legislation and recognized by the courts.

By consolidating oversight of health and safety regulation in the White House, the review process restructured regulatory policymaking in three ways. In combination, these changes helped to accomplish what employers hoped to do: equate their interests in profits with society's general interests in economic security and well-being.

To begin with, White House economic advisers became more important in social regulatory decision making. The president's top economic advisers were given the power to intervene, as "interested parties," in regulatory proceedings, and Paul MacAvoy (CEA chair under Ford), Charles Schultze (CEA chair under Carter), and Murray Weidenbaum (CEA chair under Reagan) took this opportunity. Moreover, the oversight staff was drawn from, or attached to, EOP economic agencies. Economists became active in regulatory policymaking, and critics of regulation were appointed to economic policy positions.

At the same time, it became harder for workers and consumers to influence agency policies. As agency heads became less important and economic advisers became more influential, nonbusiness interests found it harder to have their voices heard. According to a General Accounting Office (GAO) report on OMB activities under the Reagan Executive Order, OIRA became obsessed with maximizing White House discretion and minimizing public input into social regulatory policymaking. It made its decisions secretly; it did not keep written records; it refused to make its decision-making criteria public. And it engaged in extensive *ex parte* contact with affected business interests while resisting congressional scrutiny of its activities.[21]

The Reagan administration was forthright about its effort to give special access to business interests. The task force, it claimed, was supposed to serve as a court of appeal for business. Boyden Gray, Vice-President Bush's counsel and counsel to the task force, encouraged executives to take their problems to him: "If you go to the agency first," he told a Chamber of Commerce audience, "don't be too pessimistic if they can't solve the problem there. If they don't, that's what the Task Force is for."[22]

According to Weidenbaum, the exclusion of labor, consumer, and public-interest groups was one of the task force's principal attributes. It did not "have any interest group constituency to protect and defend. Its only constituency is the president and the president's program for rationalizing regulation."[23] James Miller, the task force's first director, praised its ability to resist demands for protection:

> The very existence of the task force can stiffen the back of an agency head who's being pressured by a constituent. He or she can say: "I'd like to do it for you, but there's no chance the task force members would go along—and they'd be right. The president set the principles and I've got to follow them."[24]

Finally, agencies were forced to take macroeconomic considerations into account when assessing regulatory actions. Beginning in 1974, agencies were specifically required to assess the impact of regulation on inflation. By 1981, when it was disbanded by Reagan, CWPS had filed approximately 125 analyses and statements covering regulations proposed by 10 executive branch agencies and 11 independent agencies. Although RARG acted in fewer cases, its interventions were more salient and served notice that an agency action had caught the attention of Carter's top economic advisers. Under Reagan, OMB and the task force rescinded or blocked 182 major rules in its first two years of operation. When it was disbanded in 1983, the task force claimed to have saved the economy between $15.2 and $17.2 billion in one-time capital costs and from $13.5 and $13.8 billion in recurring costs, including $137 million a year in OSHA compliance costs.[25]

The review process was by no means an unalloyed success. It was poorly coordinated at first, and the social regulatory agencies found ways around it. Under Ford, many agencies used the Inflationary Impact Statements to justify decisions made on other grounds. The CPWS reports tended to be after-the-fact critiques that delayed rather than significantly altered pending rules. Both CPWS and OMB complained about their inability to control the health and safety agencies.[26]

The Carter administration, more sympathetic to the agencies than either Republican president, created the Regulatory Council to provide the regulators with their own forum, and they used it to facilitate interagency efforts to combat economic review and justify continued rulemaking. As a result, the Carter review program worked at cross purposes. The CEA, CWPS, RARG, and OMB pressured the health and safety agencies to scale back their regulatory proposals. Through the council, agency heads coordinated their case against economic reviewers. In 1979, the heads of OSHA, EPA, CPSC, and FTC formed an additional group—the Interagency Regulatory Liaison Group—to take their case to the public.[27]

The White House reviewers were not easily deterred, and the agencies finally succumbed to pressure in the second half of Carter's term. In 1979, OMB announced that it would use compliance with the Executive Order as one criterion in reviewing agency budget requests; and in 1980, the CEA endorsed a proposal to create a "regulatory budget" that would cap the total costs that any single agency could impose on the economy. The agencies heard the message. The OSHA standard-setting and enforcement activities declined significantly after 1978. The EPA agreed to do risk–benefit analyses of its standards and experimented with incentive-based compliance methods such as the "bubble," a plan to allocate pollution "rights" to plants and firms in a designated geographical area and then allow them to trade these rights among themselves. The Reagan program, explicitly designed to overcome resistance, overwhelmed the agencies. All of the president's powers, from appointments to oversight to budgets to Court actions, were used to secure White House control over the health and safety agencies. By the end of 1984, social regulation was dead in the water.

Legal Challenges to Presidential Intervention

Presidents had previously reorganized the executive branch and attempted to increase their control over administrative agencies. Indeed, presidential calls for government reorganization are a well-established part of the American tradition of political reform. The original Bureau of the Budget grew out of White House efforts, earlier in the century, to increase its control over the bureaucracy. But the degree and kind of control achieved by the regulatory review program were unprecedented. Previous administrative reorganizations proceeded from legislative grants of authority and were subject to congressional approval. They dealt with administrative structure rather than the content of policy. Where reorganization was justified substantively, it was almost always in terms of government "efficiency" rather than the consequences of public action on the private economy.

In contrast, the White House's new powers, particularly OMB's expanded role in regulatory "management," went beyond structural considerations to the heart of substantive policy concerns. Especially after 1980, OMB's authority to oversee agency compliance with the procedural requirements of the review process was used to write, rewrite, and rescind agency rules.

Moreover, unlike previous reorganization efforts, OMB's broad powers were not firmly rooted in statute or precedent. The review program's authority derived from Executive Orders establishing it and three ancillary statutes: the 1974 Act that established CWPS; the Paperwork Reduction Act of 1980; and the 1980 Regulatory Flexibility Act, which gave OMB the power to monitor the impact of regulations on small business. These statutes were much too specific to justify OMB's comprehensive role. And the Executive Orders raised rather than resolved troubling constitutional issues, ranging from due process protections to the limits of presidential authority to revise congressional statutes through the implementation process.

The legality of OMB's role in overseeing regulation became the subject of intense political controversy as soon as the Ford administration augmented OMB's authority. Organized labor, the consumer lobby, and the public-interest and environmental movements challenged the propriety of OMB oversight, and congressional liberals protested the review program. Committee chairs sympathetic to the agencies responded to Ford's 1975 Regulatory Summit by stepping up committee oversight of the review process. Subsequently, liberal Democrats in the House and the Senate maintained constant pressure on the reviewers and supported agency efforts to avoid White House control.[28]

Both chambers held hearings on the constitutionality of OMB over-sight and its applicability to the OSH Act during the Carter adminis-tration. In keeping with the tradition of congressional deference to the executive in administrative matters, liberals in Congress were willing to accept some OMB supervision of the agencies, but they insisted that there was an important difference between "monitoring" agency compliance with an Executive Order and "assuring" the adoption of OMB recommendations. Carter's reviewers, they claimed, were en-gaged in the latter, and they challenged policies that gave OMB authority to "override or substantially interfere with or delay deci-sions on substantive issues made by legislatively designated decision makers."[29] Representative Paul Rogers (D-Fla.), chair of the House Commerce Subcommittee on Health and Environment, condemned the Regulatory Calendar as "blatantly illegal." Senator Edmund Mus-kie (D-Maine) asserted that Congress had already taken cost prob-lems into account in writing the underlying statues. "Second guessing" by "bureaucratic-economists" was inappropriate and unnecessary.[30]

Critics of the review program raised two sorts of legal challenges. The first focused on procedural issues raised by the personal interven-tion of White House reviewers into rulemaking activities. The second targeted administration interpretations of the substantive legislative grants of authority to the agencies.

The EPA, OSHA, and other social regulatory agencies make rules "informally." They can, if they choose, establish general rules based on an in-house, expert review of the available evidence pertaining to the problem at hand. They do not, for example, need to resort to a formal, trial-like procedure in which affected parties cross-examine witnesses. Nor must they deal with each hazardous situation indi-vidually. Their discretion, however, is limited. Congress, in the Ad-ministrative Procedure Act of 1946, and the courts, through judicial review of agency proceedings, have established guidelines for in-formal rulemaking. Agency officials must give advance public notice of proposed regulations; they must allow the public an opportunity to comment on proposed actions, and they must consider these com-ments; information that is not subject to public review cannot be used as a basis for their decisions; and agencies must explain and support their actions with reference to information in the public rulemaking record. Moreover, the courts have generally prohibited *ex parte* contacts—representations made outside the official rulemaking process—between agency officials and interested parties. Such contacts are held to violate the spirit of open and accountable agency deliberations based on a public record.

Nonetheless, the review program encouraged informal White House negotiations with agency officials, particularly after the close of the formal comment period, and Carter's critics argued that these were illegal ex parte contacts. In *Natural Resource Defense Council v. Schultze*, the NRDC asked the court to bar the CEA chairman from participation in the development of strip-mining regulations for this reason.[31] In *Sierra Club v. Costle*, environmentalists challenged presidential interventions in the development of new-source performance standards for steam electric power plants on similar grounds.[32]

White House oversight under Reagan raised additional questions about the legality of the review process because of OMB's extra effort to impose strict central control. A Congressional Research Service report on the effects of the Reagan Executive Order on the regulatory process summarized the principal procedural objections. First, in exercising control over final agency rules, OMB violated the separation of powers by supplanting congressional with executive authority. Second, the Executive Order violated the procedural safeguards established in the Administrative Procedures Act in several ways: it gave the director of OMB the authority to determine when an agency could issue a rule; it imposed uniform procedural standards on agencies with disparate legislative mandates; and it imposed biased decisional rules such as cost–benefit analyses on agency decision making. Finally, the review process violated the due process rights of parties affected by regulation because it failed to prevent undisclosed, ex parte contacts. In fact, it created the possibility that White House aides would serve as "conduits" for private lobbyists and/or seek to control rulemaking after the close of the formal comment period.[33]

The Executive Order's requirement that agency rules pass cost–benefit analyses was particularly problematic because it either contradicted or reinterpreted existing statutes. Most of the social regulatory laws do require the agencies to weigh multiple considerations, including the reduction of risk to health and safety and adverse economic effects, before taking action. But specific provisions vary considerably. Some statutes require that several factors be taken into account. The National Environmental Protection Act of 1969 tells agencies to consider the economy and existing technology along with the environment. The Toxic Substances Control Act requires the EPA administrator to consider the environmental, economic, and social impact of actions. Some statutes specify balancing, but do so in general terms. Under the provisions of the Consumer Product Safety Act, the CPSC's standards must weigh the risks of injury against the economic impact of standards on consumers and affected businesses. The Clean Water Act requires state water-quality standards to take into account the availability of water for industrial uses, as well as the cost of technology and the age of existing equipment and facilities.

Some statutes require cost–benefit analyses. Under the terms of The Federal Insecticide, Fungicide and Rodenticide Act, for example, the EPA must compare the economic, social, and environmental costs of prohibiting the use of particular chemicals to the attendant health benefits of regulation. The Clean Air Act requires that standards that require emission-control devices or prohibit the use of certain fuels or fuel additives must be accompanied by a cost–benefit analysis.

Still, Congress generally left the agencies with discretion to balance these multiple objectives as they saw fit. When it required cost–benefit analyses, Congress left the agency free to choose a specific decisional framework, to decide how various factors were to be defined and weighed, and to resolve other issues raised by the application of this technique to social regulation.

In contrast, the Reagan Executive Order imposed a uniform set of procedural and substantive obligations on the agencies. All agencies were required to justify rules with cost–benefit analyses; all were required to select regulatory objectives that "maximize the net benefits to society" and choose rules "involving the least net cost to society." Moreover, regulations not only had to pass cost–benefit tests but had to meet a specific definition of cost effectiveness imposed by the White House. And the order gave OMB the authority to specify precisely how the analysis should be done. The administration was sensitive to the legal issues posed by this review process—the order applied only "to the extent permitted by law." But, in practice, OMB imposed a common set of criteria on the social regulatory agencies and did not draw clear distinctions among the various health, safety, and environmental statutes. The OMB review of OSHA rules was particularly controversial because none of the provisions of the OSH Act require OSHA to take costs into account. In fact, its language and its history, as we saw, and as the Supreme Court decided in 1981, are relatively inhospitable to economic review.

Congressional Deference to the Executive

Since the New Deal, Congress has generally deferred to presidential leadership of the executive branch and confined itself to more limited oversight activities. Although congressional liberals responded critically to White House efforts to impose economic review on social regulation, this pattern of deference held here as well, and the legislature failed to curtail the growth in OMB's review powers.

The failure of Congress to block the White House review programs is striking, since the legislature had its own institutional reasons to do so. White House oversight threatened the legislature's autonomy and gave administrators the power to rewrite statutes, and institutional interests might have reinforced liberal efforts to limit economic review. They did not. Instead, political and ideological factors proved more important in determining Congress's response to White House review than any sense of its own institutional prerogatives.

Many conservative legislators, for example, seem to have bristled at the White House's effort to diminish the role of Congress. But they also wanted to deregulate industry. Some sought to reconcile these competing interests with proposals to create new forms of legislative oversight. One group sought to codify the review program and increase Congress's authority over the agencies by creating a legislative veto of agency rules. Others proposed a legislatively mandated cost-benefit test for health and safety rules.

Conservatives outside Congress warned that efforts to codify regulatory review could be counterproductive. They noted that many regulations that imposed substantial costs on business could pass cost-benefit tests. If these costs were to be controlled, the White House would have to exercise more discretion than conservative legislators might like. Anthony Scalia, the coeditor of the American Enterprise Institute's *Regulation* magazine, warned congressional advocates of a statutorily imposed cost–benefit test requirement that it could back-fire. "Regulatory reformers who do not recognize this fact," he cautioned, "and who continue to support the unmodified proposals of the past as though the fundamental game has not been altered, will be scoring points for the other team."[34] James Miller, director of the task force, told congressional staffers that an omnibus regulatory reform bill could create "impediments to the kind of work that needs to be done."[35] In response, congressional advocates of legislatively imposed cost–benefit analysis dropped their proposals and accepted the White House's version of economic review.

In contrast, liberal Democrats supported the administration position, even though they opposed deregulation. In their case, strategic considerations dominated. As their political power declined in the 1970s, they endorsed presidential oversight in order to block more conservative, legislatively imposed reforms. At the end of the decade, they joined Carter, who opposed any congressional interference with the oversight program and, together, they offered moderate, committee-supported bills to preempt conservative demands for legislative deregulation. These bills proposed a "balanced" review process that required agencies to do economic impact studies under OMB supervision, but freed them from rigidly applied cost–benefit tests.[36]

After 1980, liberals joined Reagan to block efforts by congressional conservatives to impose legislative vetos, heightened judicial review, and strict cost–benefit tests on the agencies. Reagan did not oppose congressional efforts to codify and supplement restrictions on the agencies; he supported the idea of amending the OSH Act and extending White House control over the independent commissions. But he opposed legislative measures that restricted OMB's discretion. In a marriage of convenience, Reagan allied with consumer, labor, and environmental groups to kill Senator Paul Laxalt's (R-Nev.) Omnibus Regulatory Reform bill, a proposal that would have strengthened the role of Congress at the expense of the White House's discretionary powers.[37]

The Courts and Economic Review

The courts could not avoid playing a major role in defining the limits of the White House review program. Judicial review has traditionally played an important part in structuring the regulatory policymaking process. Proponents and opponents of health and safety regulation understood this and took every opportunity to appeal agency decisions and OMB's efforts to shape them. Still, like Congress, the courts deferred to the president. The judiciary established some limits on the White House, particularly in the case of OSHA. But, overall, economic review was allowed to take root and flourish.

The courts were particularly reluctant to condemn the procedures adopted by White House reviewers. The NRDC's claim that Schultze's intervention constituted an illegal *ex parte* intervention was rejected. In this case, the court held that the availability of written records of contacts between the CEA and outside parties allowed a reviewing court to determine whether the final rule was based on information in the record or industry lobbying. The Sierra Club's challenge to White House intervention in EPA rulemaking on steam electric power plants was also dismissed. In that case, the court agreed with the White House: the president and his aides had the authority to supervise the activities of executive branch agencies, to monitor their activities, and to engage in unrecorded face-to-face contacts with regulators after the close of the period for public comments. The court acknowledged the possibility that presidential intervention might become a conduit for industry influence. But it argued that political factors were a necessary and inevitable part of the regulatory process. "We do not believe," the court of appeals wrote, "that Congress intended that the courts convert informal rulemaking into a rarefied technocratic process, unaffected by political considerations or the presence of Presidential power."[38]

When they addressed the substantive issues raised by economic review of OSHA, the courts took a more interventionist position. This, however, cut both ways. On the one hand, they interpreted the act's language to allow for economic review. At the same time, cost–benefit analysis of health standards was rejected.

The first important judicial ruling on economic review occurred in 1974, when the court of appeals resolved the controversy over the asbestos standard described in Chapter 4. The courts further limited OSHA's discretion in 1980 in the benzene case. Then the Supreme Court ruled that OSHA had to provide quantifiable evidence of the benefits of health standards. Benzene is a known carcinogen and in dealing with known carcinogens it was agency policy during the Carter administration to adopt the lowest exposure level consistent with technical feasibility and the economic survival of the entire industry. In 1978 OSHA established a 1-ppm benzene standard and rejected a petroleum industry proposal for a more lenient, 10-ppm rule. The industry challenged OSHA's standard on the grounds that it had failed to determine the dose–response relationship between benzene and leukemia, the principal health hazard with benzene. Therefore, it claimed, the agency could not specify the benefits that would result from the costly tenfold reduction involved in adopting the 1-ppm rule rather than the industry's 10-ppm recommendation.

In contrast to the 1974 asbestos ruling, the Court looked to Sec. 3(8) of the act, which generally defined a standard, rather than to Sec. 6b(5), which established the criteria for health standards. Based on this section, it imposed a "substantial evidence" test on all agency rules. Sec. 3(8) of the act, it argued, required standards to be "reasonably necessary or appropriate." That meant that OSHA had to demonstrate that its preferred exposure level was more "necessary or appropriate" than the alternatives it rejected. Specifically, it had to demonstrate that a standard eliminated a "significant risk." This was not, the Court insisted, a cost–benefit test. It was a stiff test nonetheless. Despite the absence of reliable data on it, the agency had to justify its standards with hard evidence about the relationship between levels of exposure and morbidity. This would be difficult where costs were high and health effects uncertain.[39]

Taken together, the asbestos and benzene cases forced OSHA to consider economic impact in two ways. First, the agency had to make sure that entire industries were not threatened by particular standards. Second, it had to attempt to quantify the benefits of particular exposure levels. Several issues, however, were left unresolved. Most important, these cases did not directly address industry's demands for cost–benefit analyses of OSHA standards.

This issue was settled by the 1981 cotton-dust ruling. In this case, the textile industry challenged OSHA's standard because the agency had failed to perform a cost-benefit analysis. Many observers expected the Court to find for the industry; the benzene decision appeared to foreshadow further judicial limits on OSHA's discretion. Moreover, the Reagan administration supported the industry—the standard had been issued during the Carter administration and did not reflect Reagan administration policy on regulation. In an unusual move, it urged the Court to return the standard to the labor department so that a cost-benefit test could be done.

This time the Court rejected the industry position and restricted oversight of OSHA. Here the Supreme Court returned to Sec. 6b(5) and emphasized the protective vision of the act. According to the Court, the law already contained a general congressional decision about the appropriate relationship between costs and benefits. Congress, the Court ruled, understood that worker protection would be costly and might reduce profits. Nonetheless, Congress decided that, in dealing with health standards at least, practicability was the only limiting criterion.[40]

In combination, then, these three cases simultaneously established the limits of agency rules and White House oversight. Standards had to meet the "significant risk" doctrine promulgated by the Court in the benzene case. They could not threaten the existence of entire industries, as the court of appeals had held. A standard that satisfied these two criteria was "capable of being done" or, in the language of the act, "feasible." Otherwise, OSHA was precluded from using cost-benefit tests to determine what to regulate or the appropriate levels of protection.

White House Review of OSHA

From the moment that President Ford made OSHA a prime example of overregulation, economic reviewers peppered the agency with critical commentaries on standards proposals. The sheer number of interventions into OSHA rulemaking is impressive, as Table 6.2 indicates.

Beyond their number, these interventions were important for what they said because they argued for a dramatic reinterpretation of the

Table 6.2.   Regulatory Review Interventions in OSHA Rulemaking

1975   CWPS asks OSHA to postpone noise standard.

1976   CWPS criticizes OSHA coke-oven proposal.

CWPS criticizes noise proposal.

President's Task Force on Improving OSHA Regulations established.

CWPS criticizes OSHA agricultural-sanitation proposal.

CWPS criticizes arsenic proposal.

1977   CWPS criticizes deep-sea-diver proposal.

CWPS criticizes lead proposal.

CWPS criticizes sulfur-dioxide proposal.

CWPS criticizes benzene proposal.

CWPS criticizes cotton-dust proposal.

Carter endorses economic incentives for OSHA.

Interagency Task Force on Workplace Safety and Health established.

1978   Charles Schultze, CEA chair, urges Marshall to revise cotton-dust standard

CWPS and RARG criticize acrylonitrile proposal.

CWPS and RARG criticize policy on carcinogens.

RARG reviews lead proposal.

1979   RARG reviews pesticide standard.

1980   CWPS criticizes revisions of electrical safety standards.

1981   Department of Labor withdraws labeling standard, carcinogens policy, noise standard.

Reagan asks Supreme Court to return cotton-dust and lead-standard cases for cost–benefit analysis.

1982   OMB objects to proposed labeling rule.

1983   OMB objects to proposed cotton-dust revision.

OMB recalls proposal for ethylene dibromide standard.

1984   OMB objects to proposed ethylene oxide standard.

rights created by the OSH Act. In attempting to make OSHA's health standards more efficient, White House reviewers also argued for an economist's notion of protection, and this contradicted the OSH Act's provisions. As I have noted in several places, the OSH Act created a universal right to protection. But cost–benefit and cost-effectiveness tests argued for disaggregating that right in the name of efficiency. Two rulemaking cases illustrate this view in practice.

In 1975 OSHA proposed to reduce the permissible exposure limit (PEL) for workers exposed to coke-oven emissions. Coke is a by-product of coal and is used as a fuel in steelmaking and in foundries, and as a reducing agent in blast furnaces. The materials produced during the distillation of coal are known to cause lung cancer, skin cancer, and cancer of the urinary system. In the mid 1970s, approximately 22,100 workers were believed to be at risk from this hazard. When it proposed to reduce the PEL, the agency submitted an IIS to OMB in conformance with the requirements of Ford's Executive Order. The IIS estimated that the coke-oven standard would cost between $218 million and $241 million per year and save up to 240 lives yearly.

The following year, CWPS challenged OSHA's analysis of the standard's costs and benefits, as well as the logic underlying its decision to regulate. Although OSHA had refused to do a cost–benefit analysis, CWPS did one for it. Using OSHA's own figures, CWPS calculated that OSHA proposed to "spend" between $9 and $48 million to save the life of a worker at risk. Also, CWPS supplied its own figures and did a second analysis along the same lines. It lowered OSHA's estimate of the number of lives that would be saved yearly as well as its cost estimate. Based on these new figures, CWPS calculated that the true costs of protection in this case were between $4.5 and $158 million per life. This was, CWPS suggested, excessive. It recommended that OSHA consider regulating risks in "other occupations with both higher relative risks and much larger absolute numbers" of workers at risk, thereby taking advantage of "the potential of saving more lives at lower costs."[41]

Although OSHA rejected the CWPS recommendations—it cited the act and the 1974 court decision limiting economic review to the determination that standards not endanger the viability of entire industries—the coke-oven case was an important landmark in the development of economic review. The reviewers were not content to check OSHA's estimates; they did not argue that the steel industry would be imperiled by the standard; they did not maintain that the macroeconomy would be weakened. Instead, CWPS used economic review to argue for the disaggregation of the right to protection and against the equalization of risk for workers. Efficiency considerations, CWPS maintained, argued for leaving the affected steelworkers unprotected and protecting workers in other industries where morbidity rates were high and the costs of protection low.

Actually, CWPS used the issue of equity—central to the vision of the act—against OSHA. It agreed that it "must be considered," but CWPS interpretation of equity in this context was based, not on the language and history of the act, but on an economist's reading of the problem of protection:

There are many other occupations . . . that present us with the
potential of saving more lives at lower costs. This is
important because the nation's resources that may be devoted
to saving lives are limited. The total number of lives saved
can be maximized only if the expenditures devoted toward
saving lives are made in such a way so as to be equalized
at the margin.[42]

The labeling standards case also illustrates how the review program led to the administrative reinterpretation of the worker rights in the OSH Act. This controversy began in the last days of the Carter administration when OSHA proposed a rule requiring employers and chemical manufacturers to identify chemical hazards in the workplace and make this information available to employees. In keeping with the administration's regulatory relief program, Thorne Auchter, Reagan's choice to head the agency, immediately withdrew the proposal for reconsideration. There was strong employer support for some sort of agency action, however. Supported by unions, environmentalists, and public-interest groups, the "right to know" movement was proving successful in passing state and local labeling laws; many of them were quite stringent. After 1980, the chemical industry and other affected employers shifted their position on federal regulation and sought sympathetic federal action as a means of preempting more hostile state and local regulation.

In response, Reagan's OSHA proposed a revised rule that substantially limited coverage and augmented employer rights to withhold trade secrets. In keeping with the Executive Order, the agency submitted a cost–benefit analysis that calculated the rule's benefits to be $5.2 billion versus compliance costs of $2.6 billion (both in present, i.e., discounted future, values). The rule, OSHA claimed, would save medical costs and augment labor productivity because better-informed employees would have fewer injuries and illnesses. Moreover, it would serve the public interest in two ways. State-level protection would lead to uneven coverage, whereas OSHA's rule would provide uniform protection. In addition, the labeling standard would provide workers with more information—an important but under-supplied public good.

The OMB rejected OSHA's logic and the rule. The claim by OSHA that more information and uniform standards were in the public interest did not adhere to the economic values of the review process. From an economic point of view, a federal standard was a positive disability. "By increasing smaller companies' overhead costs," OMB argued, the standard "would put them at a competitive disadvantage relative to companies that already have such programs." In addition, information could not be treated as a general public good. Although "some value should be ascribed to knowledge even if it does not improve safety," OMB admitted, "this knowledge should not be considered a 'right' in isolation from cost considerations." To OMB, the survival of small business was as important as worker protection, and workers' right to know about hazards had to be balanced against economic considerations.[43]

The Regulatory Environment

Overall, OSHA's record of wins and losses against the White House reviewers was mixed. Before 1981, the agency tried, often successfully, to resist OMB control. Under Ford, the economic review program led to criticisms of OSHA, and as the next chapter describes, delays in rulemaking. But the agency was able to avoid more serious damage. In the second half of the decade, union protests and congressional pressure deterred Carter's domestic policy advisers. And OSHA successfully resisted the reviewers' demands that it change its enforcement strategy from penalty-based inspections to incentives; it rejected CWPS's and RARG's attempts to force it to set performance standards and avoid engineering controls. The efforts by CEA chair Schultze to secure significant changes in the cotton-dust standard were personally rejected by Carter after the labor movement and Secretary of Labor Marshall came to OSHA's aid. After 1981, the review program was considerably more successful, although political pressure by industry succeeded in at least one case in overturning an OMB decision.

Nonetheless, the agency's batting average is probably less important than the overall impact of the review process on the agency's decision-making processes. On this level, the trend is quite clear, as the next chapter indicates. The centralization of executive oversight in the White House and the introduction of economic criteria not found in the OSH Act into agency rulemaking created a highly politicized and uncertain regulatory environment that undermined OSHA's autonomy and its ability to make rational health and safety policy. In this environment, OSHA could not do the things that it had to if it was to establish itself as an effective, expert agency. It could not devote its time and resources to building its organizational competence, to experimentation, and to deliberation. Instead, it was caught in a struggle of titanic proportions, one in which agency policies came to symbolize the class affiliations of the state itself. And that struggle, rather than the realities of workplace safety and health, finally determined the course of the agency's policies.

[7]
OSHA

n adopting a liberal approach to workplace safety and health, Congress left workers vulnerable to the short-run balance of political forces: protection became dependent on agency enforcement of standards, enforcement dependent on the immediate political calculations of elected officials in Congress and the White House. Indeed, the act encouraged workers to rely on a particularly narrow form of state action, factory legislation, rather than build political and economic institutions that might facilitate a more radical restructuring of the relations between employers and employees.

For the most part, the labor movement remained trapped within this framework. Despite worker interest in occupational safety and health, and the popularity of environmental protection with wider constituencies, the unions continued to concentrate their efforts on short-term economic gains rather than make working conditions a priority. In contrast, employers rose to the occasion and mounted a well-funded and well-coordinated campaign against OSHA. Concurrently, the White House's interest in promoting economic recovery dovetailed with employer concerns and led to intense scrutiny of the agency's activities.

As the political climate shifted against OSHA in the 1970s, the agency was forced onto the defensive. Business groups and sympathetic senators and representatives flooded Congress with amendments to the OSH Act and appropriations riders. Between 1971 and 1976, an average of a hundred bills a year were introduced that would have limited the agency in some way.[1] After 1976, appropriations riders exempting small business from enforcement became commonplace. Congressional oversight was also intense, peaking in the late 1970s as the business offensive against social regulation reached a crescendo.[2]

Business opposition did not preclude regulation. Oversight by the House and Senate labor committees was designed to prod the agency into action rather than restrict its activities. All substantive amendments to the OSH Act, including one that would have required cost-benefit analysis and another that would have exempted nearly 90% of workplaces from OSHA enforcement, were defeated. Carter's appointees to head the DOL and OSHA supported occupational safety and health regulation and attempted to increase the agency's standard-setting and enforcement activities. As this chapter details, however, in combination, OSHA's command-and-control approach, employer opposition, and White House review limited what any administration could accomplish.

An Overview of OSHA

Taken as a whole, OSHA's record indicates a sharp gap between the promise of the OSH Act and the agency's performance. In its initial years, OSHA expanded rapidly as it established itself as an organizational entity. The growth in budgets and personnel subsequently leveled off and, after 1980, declined (see Figures 7.1 and 7.2). The OSHA budgets actually grew more slowly than total domestic spending from 1979 to 1982. Most important, except for one short period of relatively aggressive implementation during the Carter administra-

Figure 7.1.   OSHA Budgets, 1971–1984

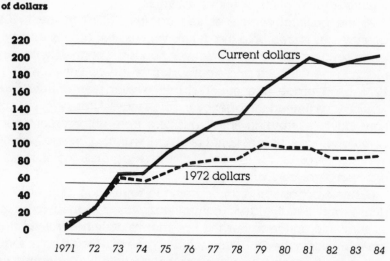

Sources: Office of Technology Assessment, *Preventing Illness and Injury* (Washington, D.C.: GPO, 1985), table 12-1, supplemented by *Economic Report of the President, 1985* (Washington, D.C.: GPO, 1985), and the OSHA Budget Office. "Budget" here is the congressional appropriation. Figures are adjusted to constant (1972) dollars using the implicit price deflator for federal government nondefense purchases of goods and services for 1972–1984 and rounded to the nearest $1000. The index for total federal government purchases of goods and services was used for 1971. See *Economic Report of the President, 1985*, Historical Tables, and table B-3 for price deflator. The appropriations for 1971 and 1972 were drawn from the budget of the Workplace Standards Administration.

tion, the agency's standard-setting and enforcement activities were extremely limited.

In its first year of operation, OSHA adopted over 4000 general industry standards and an additional set of rules covering particular industries, notably construction. Few of the general industry standards dealt with health hazards. Subpart Z of the original 1971 standards package included 400 TLVs for toxic and hazardous substances, but only 20 actually set exposure limits. In 1978 the agency revoked 982 of the original general industry standards—many had been ridiculed as "nitpicking" and "trivial"—concluding that they were, in fact, irrelevant to safety.

Apart from revising the general industry standards initially

Figure 7.2.   OSHA Staffing, 1971–1984

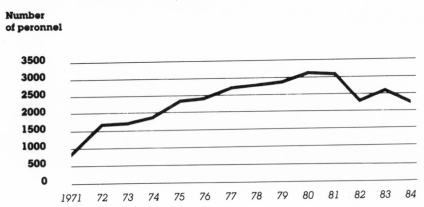

Source: Office of Technology Assessment, Preventing Illness and Injury (Washington, D.C.: GPO, 1985), table 12-1.

adopted, OSHA devoted its standards activity to rulemaking on health hazards. This proceeded slowly—the average health standard took almost three years to develop—and was quite limited.[3] The agency adopted only 15 major health rules between 1971 and 1984; Only 11 of them actually set exposure levels. Two rules, the labeling standard and the access-to-medical-records rule, dealt with worker rights to information; a third, the cancer policy, established a framework for health standard setting; and a fourth, the 14 carcinogens standard, established work practices for the safe handling of the substances (see Table 7.1). Moreover, all but two of the new health rules revised existing TLVs from the original standards package. Thus the total number of toxic and hazardous substances covered by OSHA exposure limits remained slight—less than two dozen—despite the fact that there are approximately 2000 suspected or known carcinogens in use in the workplace.

The enforcement effort was also limited in scope and intensity (see Table 7.2). The number of annual inspections increased rapidly through 1976, but was subsequently reduced by about one-third. As a result, the percentage of private-sector workers actually covered by OSHA inspections was low. In 1980, for example, inspected establishments included only 5% of nonagricultural, nonmining, private-

Table 7.1. Major New OSHA Health Standards

Year	Standard
1972	Asbestos
1974	Fourteen carcinogens
	Vinyl chloride
1976	Coke-oven emissions
1978	Benzene
	DBCP
	Arsenic
	Cotton dust
	Acrylonitrile
	Lead
1980	Cancer policy
	Employee access to medical records
1981	Noise exposure/hearing conservation
1983	Labeling
1984	Ethylene oxide

sector employees.[4] Violations and penalties were also low. Like inspections, the number of violations rose through 1976 and then declined quite dramatically to one-third or less of the peak-year total. Over a 13-year period, the average penalty per violation was $56.27. The average penalty for a serious violation—one in which the employer knows or could have known about a risk that poses a substantial probability of death or serious injury, or one in which a number of nonserious violations are grouped together—declined steadily from the agency's first years until 1983, and then rose slightly. The great majority of violations were "nonserious" or de minimis. And while the act authorized the agency to fine employers up to $1000 per serious violation, these fines averaged $263.05 between 1973 and 1985.[5]

Finally, despite the importance that health hazards played in the struggle over the OSH Act, and the threat posed to workers by them, enforcement focused on safety rather than health. The proportion of inspections devoted to health hazards never exceeded 20%. In contrast, violations for machinery and machine guarding, only one kind of safety hazard, regularly accounted for over one-quarter of all violations. Those health violations that were cited tended to be classified as less than severe.[6]

The Two Faces of OSHA

Within this basic trajectory, two distinct modes of policymaking can be distinguished, one characteristic of the agency during Republican administrations and the other characteristic of the agency under Carter. Indeed, perhaps the most striking fact about OSHA policy is the degree to which the agency responded to the changing balance of political forces by vacillating between two almost completely contradictory approaches to occupational safety and health.

In theory, administrative regulatory agencies are not supposed to act this way. They are created to provide an environment in which professionals can apply their expert knowledge to the design of rational policies. These policies can then be tested through implementation and corrections made based on feedback from the field. But OSHA did not have the opportunity to pursue this tack. Instead, it developed two competing approaches to workplace regulation and shifted back and forth between them depending on changes in its political environment. I call these approaches *voluntarism* and *activism*. This section distinguishes between them and considers their impact on standard setting and enforcement.

Voluntarism

Voluntarism has been the dominant Republican approach to workplace regulation. In effect, it has meant self-regulation by employers. In this mode, OSHA has done little to challenge corporate control over work. Nor has it sought to impose substantial costs on industry.

Table 7.2. An Overview of OSHA Enforcement

Fiscal Years

	1973	1974	1975	1976	1977	1978	1979	1980	1981	1982	1983	1984	1985
Establishment inspections (in thousands)	48.4	77.1	81.0	90.5	60.0	57.3	57.7	63.4	57.0	52.8	58.5	62.1	64.2
Percentage of inspections devoted to health	.07	.05	.07	.08	.15	.19	.19	.19	.19	.17	.18	.18	.17
Violations (in thousands)	143.5	290.2	312.9	379.8	181.5	134.1	128.2	132.0	111.4	97.1	111.7	112.1	119.7

Percentage of violations for nonserious or *de minimus* violations	98.7	98.6	97.9	96.7	86.0	71.7	66.9	63.0	68.7	75.4	74.9	71.2	69.9
Total penalties ($ millions)	3.8	6.5	7.5	11.2	9.5	14.8	16.8	17.8	10.1	5.5	6.4	7.7	9.2
Penalties per violation ($)	26	22	24	30	52	111	131	135	91	57	60	69	77
Penalties per serious violation ($)	631	576	541	545	290	285	273	255	209	195	177	185	195

Notes: These data cover the enforcement activities of federal OSHA and do not include the activities of state programs. Establishment inspections do not include "records review" inspections begun in FY 1981. The period covered by the "fiscal year" changed in 1976. The transitional quarter from July 1976 to September 1976 has been omitted for presentational purposes.
Source: U.S. Department of Labor, Occupational Safety and Health Administration, "Federal Compliance Activity Reports," various years.

Instead, within the confines of the OSH Act, the agency has sought to minimize state intervention and subordinate public authority to existing private programs. Standards have been set slowly, and industry input has played an important role in rulemaking. Enforcement has emphasized cooperation between the agency and employers. The agency has deferred to firms: even when inspection activity has been high, the agency has not relied on penalties to secure compliance with standards.

Republican administrations have differed in the degree to which they have pursued these objectives. During the Nixon–Ford administrations, OSHA was less aggressively procapitalist than under Reagan. Indeed, in the last year of the Ford term, under the leadership of Dr. Morton Corn, OSHA took several steps toward a less business-oriented approach to occupational hazards. Nonetheless, the basic principles and practices of voluntarism emerged as soon as the agency began operations in 1971, and they continued in force through 1976.

During the Nixon–Ford administrations, OSHA tried to give employers and private professional organizations a formal role in agency rulemaking. In its first major action, the agency adopted 4000 privately developed "consensus" standards despite the fact that the act allowed it to develop its own rules. Subsequently, OSHA established five Standards Advisory Committees (SACs) to maximize private input further. When the committees proved sympathetic to organized labor, they were disbanded.[7] Then OSHA turned to ANSI, nonconsensus "proprietary" groups (i.e., industry standard-setting organizations, such as the American Society for Testing and Materials), and private consultants for proposals. At the same time, NIOSH, which recommended sharp reductions in existing TLVs in several cases, was asked to refrain from proposing exposure levels.

The agency's deference to private industry compounded whatever delays might have resulted from OSHA's inexperience. Between 1971 and 1976, only 4 new permanent health standards were promulgated. The agency set 15 new safety standards, but only the mechanical-power-presses rule was a major action.

Economic feasibility and costs played a major role in the determination of PELs in most rulemaking cases. In 1972 OSHA rejected union demands for a zero or "least-detectable" exposure level for asbestos. The standard was set at 5 fibers per cubic centimeter with a target level of 2 fibers per cubic centimeter by 1976. The four-year delay was specifically designed to meet industry objections to the costs of compliance. Similarly, OSHA rejected OCAW and HRG demands for a zero PEL and/or a total ban on a set of 14 carcinogens. It promulgated a work-practices standard rather than an exposure limit. Firms were ordered to reduce worker exposure "to the maximum extent practicable consistent with continued use." Cost considerations played an important role in the 1974 proposal to revise the national consensus standard on mechanical power presses. The unions wanted a "no hands in dies" rule: machines would have to be designed so that they could not be operated unless both of the worker's hands were outside the stamping area. But OSHA concluded that this was technologically infeasible, despite evidence that compliance, though costly, was possible.[8]

Not every one of the Nixon–Ford standards was this cost sensitive. Congressional pressure, the whiff of scandal, and a public health crisis focused public attention on OSHA in 1974 and led to the first strong health rule. Organized labor expected the Nixon White House to block implementation of the OSH Act and urged its allies in Congress to pressure the DOL. After several years of dilatory tactics by OSHA, including an attempt to devolve enforcement to the states, the congressional labor committees increased their oversight of agency policy. Revelations that OSHA administrator George Guenther had offered to tailor agency policy to industry concerns in order to help the Committee to Re-Elect the President raise funds for Nixon's 1972 campaign helped labor to focus public attention on the agency.[9]

In this climate of opinion, widely publicized accounts of the discovery of three worker deaths from vinyl-chloride-induced liver cancer at a B. F. Goodrich plant had an immediate impact on OSHA policy. Once again, worker health was a crisis and required quick action; in response, the agency issued the vinyl chloride standard. Unlike the asbestos standard, this rule downplayed economic considerations and emphasized the technology-forcing aspects of the law rather than economic and technical feasibility. As described above, the plastics industry issued dire warnings about the economic consequences of the rule. But, reversing previous policy, OSHA argued that the industry could maintain production and reduce exposure by developing new methods. It adopted a 1-ppm exposure level, close to the "no-detectable level" that health and safety activists recommended.

At this point OSHA began to shift toward a more activist approach to health standards. Dr. Corn, the agency's third administrator and the first public health professional in that position, publicly rejected the use of economic review to determine exposure levels. It was, he maintained, only applicable to the determination of abatement schedules.[10] The coke-oven standard, issued in 1976, reflected this logic. OSHA rejected CWPS's advice that it perform a cost-benefit analysis on its proposal. Instead, it established a PEL at a "lowest-feasible level" determined by the performance of the most advanced coke facility in the United States.

White House review, however, vitiated Corn's efforts to shift the agency's basic approach to standards, and voluntarism prevailed. Although the coke-oven standard was eventually issued, White House scrutiny of OSHA rulemaking intensified. From late 1975 to the end of the Ford administration, OSHA tried but failed to issue a package of 13 health standards proposals, including rules for arsenic, noise, beryllium, ammonia, sodium dioxide, coke-oven emissions, lead, and cotton dust. But IIS requirements imposed by the White House forced the agency to use its scarce resources to justify each proposal in detail. The entire package was delayed, and only the coke-oven rule was issued.[11]

Voluntarism meant "cooperation" and "education" rather than penalty-based enforcement, and this philosophy was reflected in the agency's record of inspections and fines between 1971 and 1976. In general, variations across administrations in enforcement activities are more subtle than variations in standard setting (see Table 7.3). Individual measures do not always move in similar directions over time. But the outlines of the voluntarist strategy are apparent in most of the activities that measure the intensity of the agency's enforcement effort, including the size of fines, the finding and fining of serious violations and repeat violations, and the proportion of follow-up inspections in the total number of inspections.

If employers are to change their behavior, enforcement must create large and certain financial penalties for violating standards. Not only must firms be inspected, inspectors must find and fine violations and

Table 7.3.   Enforcement by Administration, 1973-1984

	Nixon-Ford (FY 1973-1976)	Carter (FY 1977-1980)	Reagan (FY 1981-1984)
Average proposed penalty per violation	$25.78	$102.27	$73.79
Serious violations:			
Percentage of inspections finding	4.6%	25.9%	26.2%
As percentage of total violations	1.5%	25.6%	27.1%
Average penalty for	$558.86	$272.73	$192.63
Repeat violations:			
Percentage of inspections finding	1.3%	3.7%	1.9%
As percentage of total violations	.7%	2.8%	1.7%
Average penalty for	$229.43	$388.74	$370.30
Follow-up inspections as percentage of total inspections	17.0%	21.3%	4.4%

Note: FY 1984 data on establishment inspections with serious violations and repeat violations cover October 1983 to March 1984.
Sources: 1984 data for percentage of establishment inspections with serious violations and repeat violations from Office of Technology Assessment, *Preventing Illness and Injury* (Washington, D.C.: Government Printing Office, 1985), table A-7. All other data from U.S. Department of Labor, Occupational Safety and Health Administration, "Federal Compliance Activity Reports," various years.

reinspect plants to assure that violations are corrected. In keeping with a penalty-based approach, the OSH Act authorized penalties of up to $1000 for a serious violation of the Act and $10,000 for willful or repeated violations. The voluntarist OSHA, however, eschewed large fines, deemphasized serious and repeat violations, and avoided follow-up inspections.

On some measures, enforcement during the Nixon-Ford years appears strong. The number of inspections and the number of violations rose rapidly through 1976. In fact, both peaked during these years. Average penalties for serous violations were historically high. Taken together, these measures suggest a deterrent approach.

These trends did not challenge the voluntarist approach to enforcement. Although there were relatively large numbers of inspections, compliance officers concentrated on trivial matters and assessed small fines. Almost all violations were nonserious or de minimus. The number of violations was large, but the average penalty per violation was low. The average penalty per serious violation was high, but almost no inspections resulted in serious violations. Finally, there were few follow-up inspections.

Activism

With the change in administration, OSHAs approach shifted dramatically. During the Carter years, under the leadership of Dr. Eula Bingham, activism replaced voluntarism. In general, Carter appointees to the social regulatory agencies were sympathetic to the demands of organized labor and the consumer and environmental movements. Several appointees, such as Joan Claybrook, a former Nader aide who was made director of the National Highway Traffic Safety Administration, were drawn directly from the public-interest movement. Dr. Bingham fit this mold. A public health professor and activist, she was strongly committed to changing the agency's direction. For her, there was still a crisis at the workplace, and OSHA had to confront it. Workers were faced, she maintained, with a "national environmental tragedy" on the job. Moreover, Congress had not created the agency to "mediate between labor and management"; OSHA was created to be an advocate of worker rights to health and safety, even at the risk of economic disruption.[12] Bingham was not completely successful in reorienting the agency. Particularly after 1978, OSHA found it difficult to set new health rules or intensify enforcement. But the shift in agency strategy is clear in standard setting and, to a somewhat lesser extent, in enforcement.

First and foremost, activism meant a much more aggressive approach to health rulemaking. Under Bingham, OSHA sought quick and dramatic reductions in PELs and broad coverage for a larger number of workers. Economic and technical feasibility were deemphasized: unless compelling evidence to the contrary was presented, the agency assumed that industries could adopt new technologies and raise the necessary capital to make needed changes. In 1978 OSHA issued six major health standards and, in four of them, set the PELs at the "lowest-feasible level." At the same time, the agency actively opposed risk-benefit and cost-benefit tests for worker health standards.[13] Perhaps most important, OSHA attempted to adopt a rulemaking policy to accelerate standard setting on carcinogens in general.

The carcinogens policy was meant to be the centerpiece of the agency's new approach to rulemaking. The Carter administration was acutely aware of the standards logjam that had developed in the last year of the Ford administration. Secretary of Labor Ray Marshall and Dr. Bingham decided to expedite rulemaking by promulgating, as rules, general policies to resolve the scientific issues that caused controversy and delay in each standard-setting case.

There are, for example, few reliable epidemiological studies of workplace illnesses. The agency must rely on evidence from laboratory tests on other mammals (rats and mice) to assess the danger that a particular substance poses to humans. But there is no obvious answer to how the agency should interpret a finding of carcinogenicity in mammals; this is a policy decision. Employers have generally opposed standards based on animal studies. Scientific uncertainty, they contend, should be resolved before the agency requires firms to invest in new control methods. The unions, in contrast, have pressed OSHA to resolve uncertainty by erring on the side of protection. The issue whether to require engineering controls or personal protective devices (PPDs) also recurred in almost every case of standard setting. Employers have preferred performance standards that allow them to rely on masks, ear plugs, and protective clothing. The unions have wanted engineering standards that control hazards at their source.

Given clear agency rules on these issues, standard setting can proceed without the often redundant testimony and time-consuming deliberations that delay it. Moreover, the resolution of legal challenges to the general rules discourages legal challenges to their application in each standards case. Both expedite implementation and encourage compliance.

With the carcinogens policy, announced in the fall of 1977, OSHA proposed to settle, through rulemaking, a number of these first principles and create a set of automatic decisional rules that could be applied to a large number of hazards. As first principles, OSHA proposed that (1) regulations require the "lowest-feasible level" of exposure because there was no "threshold" level of exposure for carcinogens; (2) agency rules extrapolate from animal studies because of the difficulties involved in doing human studies; and (3) source controls be adopted unless an employer could prove that they were technically "infeasible."

The agency then divided hazardous substances into four categories, based on their known toxicity, and developed action-forcing procedures based on this categorization. First-category substances were those whose capacity to cause cancer was established in humans or in two species of mammals, or duplicated in a single species, or for which the secretary of labor had determined that sufficient evidence existed to reach such a conclusion. For these substances, OSHA proposed to issue immediately emergency temporary standards (ETSs). These standards would be followed by permanent standards that reduced exposure to the lowest level technically feasible. Whenever possible, the agency would require that substitutes be used for these substances. The agency also reserved the right to prohibit all exposure to them.

Second-category substances were suspected carcinogens, based on unreplicated positive tests. For these, OSHA would issue permanent standards, but PELs would not be set at the lowest feasible level. Third-category substances were those for which the evidence on carcinogenicity was incomplete or inconclusive. They would be treated through conventional case-by-case rulemaking. Finally, OSHA proposed to issue alerts for, and conduct investigations on, substances not yet found in the workplace but suspected to be toxic.

The original carcinogens policy was never implemented. White House opposition and legal challenges by industry forced the agency to revise and postpone it. But OSHA did adopt a "lowest-feasible-level" policy in case-by-case rulemaking. Like Dr. Corn, Bingham argued that economic impact could not be considered when setting exposure levels.[14] Unlike Corn, Bingham was able to translate this general policy into four of the six PELs issued during her administration: benzene, DBCP, lead, and arsenic.

Bingham was also active in mobilizing support for social regulation in general. She joined Douglas Costle, head of the EPA, and other bureaucrats in promoting health and safety regulation within and outside government. In fact, because the OSH Act was more stringent than most environmental statutes, Bingham was able to take the most extreme positions when she challenged economic review. Unlike Costle, for example, she publicly rejected the appropriateness and legality of all risk–benefit and cost–benefit tests for health standards.[15] At the same time, she used agency resources to gather evidence on the economic benefits of protection and cast doubt on industry data used to calculate compliance costs. Under Bingham, OSHA hired a consulting firm to study the "implicit" social costs of injuries in order to generate data to demonstrate the cost effectiveness of regulation. It also contracted for peer reviews of cost estimates that the agency believed exaggerated the economic impact of standards proposals.[16]

As the Carter administration became more cost conscious and constrained by economic conditions and business opposition, OSHA did feel and respond to White House pressure. In fact, no new health standards were proposed after January 1978, and existing proposals were subject to strict scrutiny. In the cotton-dust and acrylonitrile cases, regulatory review led to significant concessions to industry. Although the agency was able to resist Schultze's efforts to block the cotton-dust rule, substantial concessions made by the Ford administration were retained, and these reduced compliance costs by almost 75%. With acrylonitrile, the first OSHA rule to be "RARGed," the agency chose an industry-supported PEL that was the least stringent of three feasible possibilities. The agency also retreated from the position that the carcinogens policy was not a "major rule" and, as such, did not require an economic analysis. Finally, OSHA hired its first full-time economist in 1979—a small but significant symbolic concession to the pressure for economic review.

Under Bingham, OSHA enforcement moved toward a deterrent strategy, although this change was less dramatic than the shift in standard setting. Enforcement was rationalized and focused rather than substantially expanded. The number of inspections and violations actually decreased sharply after 1976. But average penalties per violation quadrupled, and the agency explicitly adopted a policy of deemphasizing trivial and emphasizing serious violations. The percentage of inspections with serious violations quadrupled; the proportion of inspections with repeat violations nearly tripled; average penalties per repeat violation increased by 60%; and follow-up inspections rose by a quarter.

The agency also increased its efforts to target health hazards. The proportion of inspections allocated to health hazards more than doubled, from 8% in 1976 to 19% in 1979. Field inspectors were directed to look for serious health violations.[17] The level of health penalties also rose dramatically, increasing tenfold from fiscal year 1973 to fiscal year 1979. More could have been done. Health violations still constituted only 9% of total violations in 1979; health inspections only one-quarter of total inspections.[18] Nonetheless, the shift in strategy was clear.

Finally, OSHA attempted to increase worker involvement in enforcement. Two OSHA rules sought to lower the costs to workers of participation in decision making on workplace safety and health. The walkaround-pay rule required employers to compensate workers for the time they spent with OSHA inspectors. A second rule forbade employers to penalize workers who refused hazardous work. The agency also took worker complaints more seriously than it had previously. The proportion of inspections devoted to these complaints tripled from the Ford to the Carter administration.[19]

Undoubtedly, business and White House pressure limited the agency's ability to expand its enforcement efforts further. By the late 1970s, opposition to OSHA fines and inspections was intense. Employers were angered by the deterrent approach and sought relief from OSHRC, Congress, and the White House. Companies flooded OSHRC with appeals, and the percentage of contested inspections more than doubled.[20] And OSHRC proved sympathetic to employer complaints and regularly reduced fines and modified agency abatement orders. In some cases, it used employer challenges to enforcement orders to question OSHA's standard-setting policies.[21]

At the same time, the business lobby succeeded in securing serious congressional and White House consideration of reforms in OSHA's enforcement program. Sponsored by a coalition of Senate Labor Committee Democrats and Republicans, including many of the agency's original supporters, S. 2153, the Occupational Safety and Health Improvement bill, proposed a major statutory revision of the OSH Act. The bill's Democratic supporters argued that it would preclude more radical cutbacks. But S. 2153 incorporated most of the principal recommendations of business about agency enforcement: it exempted "safe" workplaces from inspection; eliminated most civil penalties for firms with joint labor-management safety committees, advisory consultation programs, and good safety records; and restricted the right to request inspections in exempt firms. Under these provisions, nearly 90% of workplaces would have been exempt from OSHA enforcement.[22] The White House, in turn, established an Interagency Task Force on Workplace Safety and Health to consider alternatives to OSHA inspections, including injury taxes, performance standards, and workers' compensation reform. These employer efforts were only partially successful; Congress rejected S. 2153 and the task force endorsed higher penalties rather than deregulation.[23] But heightened scrutiny undoubtedly discouraged OSHA from intensifying its enforcement effort after 1978.

The Restoration of Private Control

With Reagan in the White House, OSHA shifted direction again, and the voluntarist approach was restored. But, under Reagan, voluntarism was elevated from a pragmatic response to industry opposition to a philosophy of state action. Thorne Auchter, a Florida construction industry executive appointed by Reagan to head OSHA, withdrew booklets on cotton dust, acrylonitrile, health and safety rights, and vinyl chloride because they were too one-sided. A prominent public health activist in NIOSH was fired. The Department of Labor intervened in the pending cotton-dust and lead cases on the side of employers and asked the Supreme Court to return these standards to OSHA so that cost-benefit tests could be performed.

In keeping with the Reagan embargo on new regulations and OMB opposition to imposing substantial costs on the private sector, standard setting was first suspended, then carefully circumscribed. Over a hundred long-delayed rulemaking projects were dropped in 1981, without cost-benefit or cost-effectiveness tests to determine whether or not they could pass the requirements of the new Executive Order. Eight standards proposals were recalled or weakened.

Agency policy on costs and feasibility was reversed. Auchter dropped Bingham's "lowest feasible level" approach to standards: costs became the single most important criterion in standards actions. Although the Supreme Court rejected cost–benefit analysis for Sec. 6b(5) standards in 1981, the agency developed a stringent, fourfold cost-effectiveness test for health standards:

1. Did the hazard pose a "significant risk" to workers?
2. Did the rule substantially reduce the risk?
3. Did the standard adopt the most technologically and economically feasible approach?
4. Was the standard cost effective?

To be issued, health standards had clearly to reduce obvious hazards and to do so with only the *most*-feasible (i.e., least-costly) control technologies available.[24]

All nonhealth standards had to meet the requirements of Executive Order 12291 and provide "potential net benefits" greater than their "potential net costs." This category was interpreted broadly. Labeling and noise, or "hearing-conservation," proposals, for example, were classified as nonhealth standards; cost–benefit analyses were done in both cases.

Some standards were issued by OSHA, but these proposals were either weak and designed to preempt the possibility of stricter regulation by others or issued in the face of legal challenges. The labeling standard, for example, addressed business apprehension about the worker right-to-know movement. Compared to stringent state and local laws, the OSHA standard covered only 25% of the workers who handled toxic substances and chemicals. But because it was a federal standard, the agency maintained that it preempted stronger state and local efforts.

Other standard-setting activity was designed to weaken existing rules. The hearing-conservation standard was reconsidered and qualified. Similarly, OSHA proposed to revise the cotton-dust standard in order to remove some industries from coverage, delay compliance for others, and allow some firms to substitute respirators for more expensive engineering controls. In addition, the agency proposed to modify and dilute the rule granting employees access to their medical records and the carcinogens policy.

The only two significant new rulemaking actions were done under duress. The ethylene oxide standard was proposed after a ruling by a court of appeals forced the agency to act. Similarly, a 1984 court ruling led OSHA to propose a long-delayed standard for farmworker sanitation. In this case, the agency's reluctance was apparent in its written explanation for the rule. The agency adopted restrictive assumptions about the risks that farmworkers faced from unsanitary field conditions and the benefits to be gained from regulation. As a result, the justification for the standard was so weak in comparison to the agency's own criteria for rulemaking that it all but invited successful industry suits.

In the 1980 campaign Reagan had called for a new approach to workplace safety and health in which OSHA acted as industry's sympathetic adviser. In keeping with that laissez-faire vision, enforcement was all but completely deregulated. Average penalties per violation and per serious violation declined dramatically. Follow-up inspections declined by over three-quarters. The proportion of inspections with repeat violations dropped by nearly 40%. Perhaps employers were cleaning up on their own, but agency policy seems to be a more compelling explanation for the declining enforcement effort. Auchter self-consciously altered OSHA's approach to enforcement. The agency was not mandated, he maintained, to "intrude" in workplace issues. It was to "assist" labor and management.[25]

In practice, the agency assisted management rather than labor. Most important, it adopted administratively the inspection strategy that had failed as an amendment to the act in 1980. Beginning in 1981, OSHA exempted all but the most hazardous work sites in the most hazardous industries from routine inspection. Pilot programs were started to allow "safe" firms to avoid general inspections entirely by setting up labor–management committees. No effort was made to distinguish between committees that empowered workers and those that left all decisions with management. In fact, firms without labor–management committees could qualify for an exemption if they undertook "intensive" management programs. Concurrently, resources were shifted from on-site inspections to off-site consultation with employers. Worker-complaint inspections were discouraged by routinely referring worker complaints back to employers unless workers alleged, and established in writing and to the agency's satisfaction, that "violations threatening physical harm or an imminent danger" existed.[26] Predictably, the proportion of inspections in response to employee initiatives plummeted from a yearly average of 32.3% during the Carter administration to 14.5% under Reagan.[27] Finally, the agency directed its field offices to emphasize settlement agreements with firms and to plea-bargain citations. As employers had recommended, regional managers who failed to reduce the number of citations contested by employers were threatened with disciplinary action by their Washington supervisors.[28]

Some of the changes adopted under Bingham were continued. The share of serious violations in the total number of violations remained constant. The average penalty per repeat violation showed little change after 1981. But the agency failed to take the other, related actions that were necessary to sustain a deterrent approach to enforcement. Thus, while the share of serious violations remained high under Auchter, the average penalty for them declined. While the average penalty per repeat violation remained constant, the share of repeat violations in the total number of violations dropped significantly. The basic pattern signaled by Reagan in 1980 and endorsed by Auchter held. As an enforcement agency, OSHA was rendered toothless. It became, instead, an advocate for the employer's point of view on occupational hazards.

The Anatomy of Policy Failure

Voluntarism and activism were politically expedient responses to changes in the agency's political environment, but both approaches failed to confront the problem of occupational hazards. Voluntarism was completely unsuited to changing those aspects of work and the labor process that made working conditions unsafe. Activism took several steps in the right direction, but this approach was insufficient to improve occupational safety and health.

The analysis in Chapter 1 suggests that an effective occupational safety and health program must force employers to increase their investments in prevention and involve workers in plant governance. To accomplish these goals, the state must set standards that provide employers with clear and consistent signals about the appropriate levels and kinds of investment in protection. It is particularly important that standards anticipate hazards and that standards are revised regularly. Capital expenditures can be minimized if employers take health and safety into account when they purchase new equipment and design new plants. Capital goods producers are likely to comply with existing standards, and firms are likely to acquire state-of-the-art equipment routinely as they invest in new capital. In contrast, it is very costly to retrofit existing plants.

An adequate enforcement policy must create deterrents to the violation of standards. This can be done in one or both of two ways. First, the state can use fines to discourage employers from violating standards. If this approach is chosen, penalties must be stiff, inspections comprehensive, and policy consistent. The rational employer who is out of compliance and contemplating investing in health and safety should consider (1) the odds of being inspected each year; (2) the probability that once he or she is inspected, the violation will be detected; and (3) the size of the penalty that will be assessed if the employer's violation is cited.[29] If the odds on inspection are low, the penalties minimal, and/or policy too erratic to be predictable, the economically rational employer would be advised to violate standards with high compliance costs.

Alternatively, the state can create and subsidize a variety of in-plant mechanisms through which workers can participate in a decentralized enforcement effort, such as mandatory health and safety committees, safety representatives, and employee-run occupational health clinics. Field inspections can then be linked to worker participation in in-plant institutions to take advantage of worker knowledge, maintain motivation, and facilitate involvement in decision making over the work environment.

As the record of OSHA policy reveals, the agency failed to adopt programs of either sort. Instead, in its search for an acceptable response to the competing political pressures on it, OSHA developed inconstant and inconsistent programs that resulted in few standards and a poorly conceived and implemented enforcement program.

Even on the most conventional performance criteria, voluntarism must be judged a failure. It made no attempt to help employers or employees mount the sustained private effort necessary to monitor and prevent hazards. In defending the corporate sector, voluntarism misunderstood and distorted the realities of private power in the workplace. It failed to confront the incentives to workers and unions to trade away protection for more narrowly defined economic benefits, jobs, and organizational security. It ignored the disincentives to employers to invest in prevention, and their vested interest in preventing workers from playing a larger role in the determination of working conditions.

Activism confronted some of these issues but was overly statist and insufficiently attentive to the underlying problems it attempted to solve. Despite a few efforts to facilitate worker participation in agency enforcement, government relied on penalty-based inspections rather than encourage worker control over decisions about the conditions of work. At the same time, the level of fines and inspections remained too low to provide a real deterrent to employers.

As a result, OSHA had the worst of both worlds. It encouraged business opposition, but failed to change employer practices substantially. The agency was by no means wholly responsible for this outcome. As I stated above, organized labor bears some of the responsibility for failing to mount a concerted attack on occupational hazards or to address the problems inherent in its economistic approach to politics and the workplace. As a result, OSHA was left even more vulnerable to countermobilization by business, and more dependent on penalty-based inspections, than it would otherwise have been. This, in turn, created a vicious cycle in which business opposition made it impossible to mount the kind of penalty-based enforcement policy that could have changed employers' behavior, while the failure to improve worker health and safety legitimated business opposition.

The sheer inconsistency of agency policy compounded the problems inherent in both these approaches. Policy at OSHA failed to communicate clear signals to employers and employees about the costs of violating the law or the possibility of using the agency to facilitate worker action. Employers did not face certain costs of punishment; workers did not get consistent information about the hazards they faced or certain rights to act on their own. Instead, the wide swings in policy probably discouraged workers from organizing around the issue and encouraged employers to risk violating agency standards.

The overall effects of these two regulatory strategies, and the agency's movement back and forth between them, can be seen in the quality of OSHA's standard-setting and enforcement programs. Standard setting was particularly vulnerable to political pressure. The tug-of-war over the meaning of the OSH Act undermined OSHA's efforts to settle the question of economic and technical feasibility and establish a general standards policy. As a result, the agency could not establish a credible presence as a standard-setting organization. Few new standards were issued; outdated consensus standards were left unrevised; standards that were proposed or issued by one administration were recalled and reconsidered by the next. The one major attempt to expedite standard setting, the carcinogens policy, fell victim to White House and business pressure and was not implemented. In 1981 the agency deleted key elements of the rule that applied to first-category substances. In 1982 it proposed to reconsider the entire rule. In 1983 it stayed several of the remaining provisions. In toto, these changes nullified the policy. OSHA was back where it had started—using the slow and cumbersome case-by-case approach that had initially frustrated standard setting.

In 1985 the Office of Technology Assessment (OTA) evaluated OSHA's standard-setting effort by comparing its standards for 123 specific substances to NIOSH recommended standards and the TLVs of the ACGIH. The OTA findings suggest the inadequacy of OSHA's case-by-case approach. In the vast majority of cases, NIOSH recommendations and ACGIH TLVs were more stringent than OSHA standards. This was true for ceiling limits (maximum allowable exposure at a particular time) and "time weighted averages." In addition, ACGIH TLVs covered 200 substances that OSHA had failed to regulate. Most telling, despite continual updating by ACGIH, the PELs for nearly all of the 410 substances regulated by OSHA had not been revised since they were initially adopted in 1971.[30]

Given the need to send clear signals to employers, OSHA's failure to establish a clear and consistent policy on feasibility was particularly costly for worker health and safety. As we saw, the courts intervened three times to set the basic contours of agency policy on this issue. Judicial rulings left room for considerable discretion, however, and OSHA filled in this space in several different ways. Feasibility weighed heavily in OSHA's first two permanent health standards: asbestos and 14 carcinogens. Cost considerations played an important role in the 1974 proposal to revise the national consensus standard on mechanical power presses. The vinyl chloride standard inaugurated a new approach that emphasized the technology-forcing aspects of the act, and the coke-oven standard followed this logic. Four of the six health standards issued between 1977 and 1981 built on this strategy and adopted a "lowest-feasible-level" approach. Then, under Auchter, OSHA dropped the "lowest-feasible-level" route altogether and adopted a complex set of tests that privileged economic effects in determining what to regulate and the content of standards.

Even Reagan's approach did not produce consistent policy. As critics of economic review have maintained, these economic tests leave considerable room for interpretation. The development of OSHA's labeling standard illustrates the point. After OSHA revised the Bingham rule to limit its coverage and the rights of workers exposed to dangerous chemicals, it resubmitted the standard to OMB with a cost–benefit test that estimated the standard's benefits at $5.2 billion, or twice as much as the estimated $2.6 billion in compliance costs. When OMB rejected OSHA's analysis, it argued that the agency had exaggerated the benefits of labeling by a factor of 80 and that the standard's true benefits were closer to $65 million. When Auchter appealed OMB's decision to Vice-President Bush and the Task Force on Regulatory Relief, a third cost–benefit test, to be prepared by Professor Viscusi, was ordered. Using a different set of assumptions about worker behavior and a willingness-to-pay approach to estimate benefits, Viscusi concluded that a labeling standard would yield $2.85 billion in benefits, just enough to cover the estimated costs. Using Viscusi's study, Bush overrode OMB and approved OSHA's proposal; a final labeling standard was adopted in 1983.[31]

On the cotton-dust standard, a different kind of industry interest frustrated OMB's attempt to oversee OSHA. In line with White House proposals stretching back to the Ford administration, Reagan's OMB suggested that it would be "cost effective" to revise the Carter standard, already upheld by the Supreme Court, to allow firms to substitute dust masks for engineering controls. This change, it argued, would not violate the prohibition on cost–benefit tests of health standards; it would simply make the standard more cost effective. Given OMB's assumption that masks were as effective as engineering controls, it was able to justify this proposal in strictly economic terms. But many of the larger companies in the industry had undertaken expensive capital investment programs and were now in compliance with the 1978 standard. They wanted their competitors to match these investments. As a result, OSHA took their case to David Stockman, director of OMB, and he personally overruled OIRA—despite the obvious economic case for revision.[32]

With regard to enforcement, political opposition undermined OSHA's ability to induce compliance with those standards that did exist. The inspectorate was small, and the odds of any single firm being inspected were low. In combination, federal and state inspectors were able to inspect less than 4% of firms in a given year.[33] The average firm's chances of being inspected were approximately one in a hundred. To make matters worse, many hazards were likely to go undiscovered by inspectors: the average inspection cited 2.1 violations. The violations that were cited incurred small penalties, particularly in comparison to the costs of compliance for many standards. The average penalty during the 1970s, the heyday of agency action, was $193. Penalties for health hazards were higher and averaged a little over $400 per violation over the same period, but few of these violations were cited. The average firm was likely to be fined a total of $7.08, or 34 cents per worker covered by the act. Thus, even OSHA's most punitive actions had limited financial consequences for employers. The financial incentives for compliance were, as Viscusi concluded, "virtually nonexistent."[34]

OSHA's Impact

The OSHA record can also be gauged by considering the extent to which the agency reduced the incidence of workplace accidents, injuries, and disease. The evidence is ambiguous, but, taken together, the available studies suggest that, at best, OSHA had a small positive impact on worker health and safety.

Before examining the available data in detail, three problems that arise in interpreting the data should be addressed. First, as I noted earlier, it is difficult to measure the impact of health regulation because it is difficult to identify and explain many occupational

Figure 7.3. Occupational Injury Rates, 1972–1983

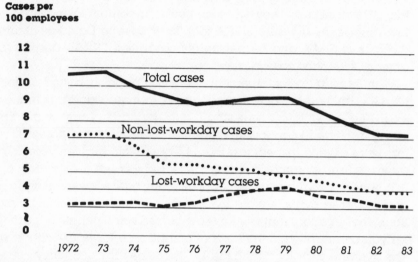

**Cases per
100 employees**

Source: Office of Technology Assessment, *Preventing Illness and Injury* (Washington, D.C.: GPO, 1985), figure 2.1.

diseases. Second, the Bureau of Labor Statistics (BLS) collects the data but has not made plant-by-plant data available to OSHA or the general public. Instead, the BLS publishes industrywide aggregate figures on injuries and illnesses for broad risk categories.[35] Data on individual companies are available in a few cases; but these cannot be broken down by plant and facilities. As a result, safe plants are grouped together with high-risk plants.[36] This makes it almost impossible for the analyst to examine the relationship between specific kinds of work sites and processes and worker health and safety.

To complicate matters, the BLS changed the definitions of workplace injuries and accidents in 1971 so that pre- and post-1971 data are not comparable. It is possible to splice the two series, but close examination of trends across this divide is difficult, making pre- and post-OSHA comparisons problematic.[37]

With these caveats in mind, consider the available evidence. At the most general level, there have been significant declines in aggregate occupational injury rates since the creation of OSHA. But as the OTA

Figure 7.4.   Injury Rate and Unemployment, Private Sector, 1972–1983

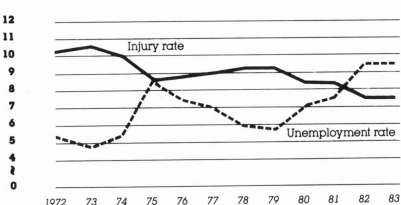

Note: The injury rate is per 100 employees.
Source: Office of Technology Assessment, Preventing Illness and Injury (Washington, D.C.: GPO, 1985), figure 2.2.

has demonstrated, these improvements are readily explained by factors other than regulatory activity.[38] Figure 7.3 charts three separate measures of the injury rate: non-lost-workday cases, lost-workday cases, and total cases (a combination of the first two measures).

The decline in total cases and non-lost-workday cases is significant. Business-cycle factors seem to account for it, however, as Figure 7.4 indicates. Injury rates rise and fall with employment rates. Other studies confirm these findings. The explanation is straightforward; it is the same one encountered in the analysis of the impact of the business cycle on the rise in accident rates in the mid 1960s. As production increases and employment rises, firms hire young and inexperienced workers. They tend to have higher injury rates than older, more experienced workers. Moreover, firms expand facilities, introduce new machinery, and increase the pace of production. These factors increase the likelihood of accidents and injuries. During recessions, the cycle is reversed. In short, accident rates declined in the early Reagan years because of recession, not voluntarism, as the administration proclaimed.

The impact of OSHA on accidents and injuries can be assessed in a variety of ways, ranging from statistical models that predict accident and injury rates without regulation to comparisons of the injury rates of inspected and uninspected firms to comparisons of the injury rates of firms inspected at different points in time.

Generalizations are difficult because the findings are mixed. Smith found that OSHA inspections lowered injury rates in small firms in 1973 but not in 1974.[39] McCaffrey found no evidence of a positive impact on safety in 1976, 1977, and 1978.[40] In contrast, Cooke and Gautschi's intensive study of one state's experience found that, between 1970 and 1976, inspected firms were more likely to experience decreases in injury rates than uninspected firms.[41] Mendeloff's analysis of California data suggested that some kinds of accidents are more easily prevented than others; OSHA helped reduce the frequency of accidents in which workers were caught in or between machinery or hurt in explosions.[42] Still, if there is a single conclusion, it is this: OSHA had little impact on safety in the aggregate and a small, positive effect in some situations.[43]

For the reasons stated, the analysis of health impact is even more problematic. As a result, there are few careful analyses, and the existing estimates vary greatly. There is a consensus that estimates of the incidence of occupational disease based on workers' compensation claims are much too low, but little agreement about the true range. One comprehensive review of the available studies found that published figures range from 10,000 to 210,000 deaths annually.[44] Three government reports are widely cited in the literature and the press: the Public Health Service's estimate that there are 390,000 new cases of occupational disease annually; HEW's finding that at least 20% of all cancer deaths are attributable to work-related exposures; and NIOSH's conclusion that 100,000 Americans die every year from occupational health hazards. These reports may be exaggerated. The OTA concludes that the evidence suggests only 5% of cancer deaths can readily be attributed to hazardous work.[45] In truth, we simply do not know enough to reach any firm conclusions.

Because of this wide variation in estimates, statistical explanations of health trends are not likely to prove rewarding, and few experts attempt them. As an alternative, some analysts have measured trends in exposure levels for selected substances. Again, the evidence is mixed and generalizations must be tentative. Mendeloff found a significant drop in worker exposure to asbestos—where OSHA had significantly lowered the preexisting threshold limit value—between 1973 and 1979, but no change in exposure to lead, silica, or tricholorethylene, where the relatively weak ACGIH standards had remained in force.[46] Other studies have found significant declines in worker exposure to OSHA-regulated substances including cotton dust, vinyl chloride, and lead.[47]

In sum, it is unlikely that OSHA had a major effect on the safety of American workers between 1971 and 1984. The agency seems to have reduced the risks of occupational disease for workers in a few industries, such as textiles, and helped to prevent some kinds of accidents, such as explosions. Overall, however, OSHA's standard-setting and enforcement activities must be judged a failure, in relation both to the hazards that workers face and to the goals of the OSH Act.

Liberalism at Work: An Empirical Assessment

Why did OSHA fail? The changing balance of political forces clearly frustrated worker protection. As economic conditions declined and business groups mobilized, neither Congress nor the White House was willing to fund or support aggressive implementation. But as I stated earlier, the liberal form of state intervention played a critical role in translating economic decline and business mobilization into effective political opposition to workplace regulation. Moreover, the characteristic infirmities of factory legislation compounded the problems created by the shift to the right. Four characteristics of the liberal approach were most important in this context.

First, by vesting responsibility for changing employer practices in an executive agency instead of attempting to devolve power to workers at the plant level, Congress left the program vulnerable to changes in the short-run balance of political forces. Economic crisis and employer opposition took such a heavy toll on workplace regulation because this approach left protection so dependent on the political interests of elected officials in Congress and the White House.

Second, despite the nominal presence of public authority in the workplace, employers remained free to organize work as they saw fit while OSHA responded to, rather than initiated, change. Most important, the agency could not adopt the long-term, coordinated policy approach advocated by the Frye Report, an approach that could have prevented hazards efficiently by anticipating changes in technology and workplace organization, integrating worker health into a larger program to improve public health, and directing investment to those economic activities that maximized employee health and well-being.

Third, OSHA's reliance on a penalty-based approach to compliance actually encouraged business opposition and facilitated White House review. The "intrusive" nature of this approach to enforcement was all but guaranteed to alienate employers. At the same time, the inherent inefficiencies of command-and-control regulation provided the agency's critics with more than enough ammunition to challenge the entire program. The most responsible observers were careful to distinguish between this approach to enforcement and federal regulation in general. But as the political climate shifted against OSHA, fewer and fewer of the agency's opponents took the time or effort to make this distinction.

Finally, while state intervention in the workplace alienated employers and facilitated business opposition, it did not encourage worker mobilization. To the contrary, the act encouraged workers to rely on state action rather than build political and economic institutions that might facilitate a more radical restructuring of the relations between employers and employees.

[8]
Regulatory Reform

T he problems facing OSHA have been severe, and, based on the record, it is clear that the American approach to occupational hazards should be changed. The preceding analysis suggests that regulatory reform must confront how the political and economic power of business has combined with an overly statist approach to regulation to limit severely the ability of public officials to change what employers do about working conditions. Too few resources have been transferred from other, more profitable investments to worker health and safety. Workers have not taken a large part in decision making over the conditions of work.

Accordingly, any serious regulatory-reform proposal will have simultaneously to increase the incentives for employers to invest in health and safety and increase workers' ability to participate in determining the conditions of work. This chapter evaluates alternative regulatory-reform proposals according to these two criteria. Three alternative proposals to restructure OSHA are examined:

the first two are conventional and have been discussed in earlier chapters. One approach is conservative; it leaves the allocation of risk and protection to private action in markets, supplemented by lawsuits. The second approach is a neoliberal one that seeks to "rationalize" administrative regulation by imposing cost–benefit and cost-effectiveness tests on standards and enforcement. Both approaches will be considered in greater detail here than was possible earlier.

The third approach is not conventional, at least not in discussions of regulatory reform in the United States. For lack of a better term, I call it a *labor-oriented approach*. This term is used to designate that cluster of features that underlies the neocorporatist and social democratic occupational safety and health programs that are common in many other advanced capitalist societies. I focus in particular on the possibilities and limits of the social democratic variant. This approach relies on a more radical, worker-oriented reordering of the relationship among the state, the firm, and the market than is usual in discussions of American policy, or likely in the foreseeable future. Nonetheless, the approach offers an important comparative perspective on the nature of the liberal approach to workplace regulation and the limits of the current policy debate in the United States.

Market-Conservative Approaches

As I noted in the discussions of deregulation and regulatory reform in Chapters 4 and 6, market-conservative approaches are based on a neoclassical economic model in which the private, self-interested actions of employers and employees in markets determine the level of protection and the choice of control technologies. Based on this model, regulatory reformers propose two changes in the current approach to occupational safety and health that could be adopted independently or used to supplement each other. The first would leave worker protection entirely to labor markets. As Smith writes, "Occupational safety would be bought and sold on the same basis as most other goods."[1] The second change would amend the current no-fault workers' compensation to facilitate employee negligence suits. This would, it is claimed, create additional economic incentives for employers to change their health and safety practices.[2]

Labor Markets and Social Welfare

The neoclassical view of the market in general, and labor markets in particular, treats occupational safety and health as a good that is bought and sold as part of the wage bargain between employers and employees. According to this model, when labor markets are allowed to operate freely, the forces of supply and demand determine the levels and kinds of risk faced by individual workers, the level of investment in occupational health and safety, and the approaches taken to control hazards.[3]

On the supply side, employers offer safe and healthy work when they offer jobs. Some employers, usually those for whom it is relatively easy to reduce risks, seek to attract workers by offering relatively safe jobs with relatively low wages. Other employers bid for workers by compensating for high risks with high wages. Given the number of different industries and occupations, and the variable costs of protection across jobs, in combination employers are likely to offer a wide variety of jobs combining different levels of risk and wages.

On the demand side, different workers have different preferences for income and safety. Some workers prefer safe work and will accept lower wages to get it. Other workers will take hazardous work if there are risk premiums associated with it. In an analogous fashion to what employers do, workers distribute themselves among risky jobs with high wages and lower-paying jobs that involve less hazardous work.

According to its advocates, the neoclassical approach is both efficient and democratic because it maximizes individual liberty and, simultaneously, leads to a socially optimal level of risk and protection. Based on information about hazards and wages, and workers' own preferences for safe work versus higher incomes, some workers will choose wages over safety and others will choose safety over wages. At the same time, competition among employers for workers will lead firms to reduce risks in the least costly manner possible. Employers who choose to offer safe jobs will also be in a position to discover the least costly methods of protection because they understand the technical details of their enterprises.

In an equilibrium situation—one in which the demand for, and supply of, health and safety intersect—the worker's total compensation (wages plus the marginal cost of safety per worker) will be equal to the value of the worker's marginal product. At that point, the supply of safety will reflect workers' demands for protection and the resource costs of reducing risk. In economists' terms, this is a *Pareto efficient* outcome. That is, no one can be made better off by a change in the level of safety without imposing costs that are larger than the protected worker is willing to pay.[4] In contrast, government action is likely to yield "suboptimal" levels of protection. State agencies will impose stricter standards than workers would themselves choose. Moreover, given the agencies' legalistic approach to social problems, they will probably prefer uniform standards and engineering controls to variable standards and PPDs. As a result, employers will be prevented from adopting the least-cost methods of controlling hazards.

Those who argue for this approach acknowledge two defects in the market for worker health and safety and recommend remedies for them. First, workers do not have accurate information about health hazards because of their uncertainty about what they are exposed to and the causes of diseases. Second, the workers' compensation system distorts incentives to firms to invest in safety because it allows employers to pool their risks. Most employers pay premiums based on industrywide injury experiences rather than their own plant records. Because insurance premiums do not accurately reflect the individual employer's efforts, the workers' compensation system prevents the firm that invests in safety from realizing the complete economic benefits of that investment.[5]

Presumably, state action can supplement the market-based allocation of risk and protection by correcting these defects. Specific changes have been proposed to solve these problems. Deficiencies in health information can be remedied by labeling laws and information programs that encourage and/or subsidize full disclosure of health hazards. Alternatively, the liability laws can be changed so that workers can sue employers to recover the costs of work-related disease. Employers could be allowed to defend themselves against charges of negligence by undertaking programs to monitor workplace exposure, disseminate current information to employees, and subsidize private research efforts.[6] Disincentives to employers to invest in safety could be eliminated by experience-rating all firms. Alternatively, small deductibles, to be paid by the employer, could provide an incentive for all firms to reduce injuries.[7]

This model looks particularly attractive when it is compared to the practical problems associated with command-and-control regulation. In the market-conservative approach, standard setting and enforcement are decentralized and, therefore, more flexible and efficient. Labor markets allow employers to choose to supply a level of safety appropriate to the costs of hazard prevention in their firms. Employers can then attempt to minimize their costs by experimenting with control methods and selecting those methods that are most suitable for particular plants. Court action targets specific employers who create unusually hazardous conditions. Negotiated court settlements encourage the development of flexible solutions to particular problems.

To advocates of this approach, markets in health and safety are not just theoretical possibilities; they exist in the real world. Reviewing the evidence on whether workers and employers act in ways consistent with this theory, Smith concludes that "a market for job safety probably does exist." Moreover, "This market functions, on the whole, as it should."[8] Some advocates of market-based incentives argue that the existence of this market mechanism allows for the dismantling of OSHA and the restoration of the private system that dominated health and safety before 1970, although it would be supplemented by a reformed workers' compensation program.[9]

A market-based approach will, it is claimed, maximize individual liberty because it is self-enforcing. Private action in labor markets replaces the bureaucratic supervision of employer practices. When state action is necessary, the courts replace the executive branch. Judicial decision making is preferred because it rests on the initiative of private parties. In either event, individual freedom is maximized.

In sum, private action seems to provide occupational safety and health effectively and democratically. If workers choose protection over wages, the level of investment in risk reduction should rise. Successful employee liability suits should reinforce the economic incentives to employers to devote resources to health and safety. Workers can even purchase, or bargain for, control over working conditions. Indeed, the threat of negligence suits might lead employers to share power with employees in order to limit their legal liability. Legislative reforms could sanction this arrangement.

The Limits of Market Capitalism

The market capitalist view must be rejected on two related grounds. First, it fails to acknowledge how markets actually work in capitalist economies. Second, it does not consider how market "imperfections" discourage investment in health and safety and worker participation in plant governance.

Consider, first, the problem of market competition. Even if we take the view that there are markets in occupational safety and health—a position that is itself controversial—it does not follow that these markets are competitive. Labor markets are imperfect, and since protection is bought and sold in labor markets—as an aspect of labor rather than a separately produced commodity—occupational safety and health markets are also imperfect. For example, discrimination and other employer practices divide workers into distinct, noncompeting groups. In addition, large firms in concentrated industries often enjoy considerable monopsony power. As a result, the tastes of the marginal worker (i.e., the last worker hired) will determine the amount of safety provided by the firm. But these tastes are not likely to represent the tastes of all workers; workers who readily move between jobs are likely to be younger and less experienced, less familiar with and worried about job hazards. As a result, the market value of protection will not reflect the preferences of workers as a whole.[10]

The chronic oversupply of labor also shapes workers' preferences for safety and health. The shortage of jobs and the fear of economic insecurity should discourage many workers from complaining about hazards and encourage them to accept higher risks than they would otherwise. The costs of moving between jobs are likely to contribute to the undervaluation of health and safety.

Another problem arises because both forms of private action—the sale of labor in markets and court suits—assume the existing class-based distribution of income and wealth. A worker's initial market position is apt to shape his or her preferences about the tradeoff between safety and wages, ability to choose among jobs, and ability to mount and sustain legal action. A distribution of occupational safety and health that reflects these inequalities should result. Thus, private action will leave poorer workers with less protection than more affluent workers enjoy.

Advocates of market-based incentive systems are generally indifferent to this outcome. Smith, the leading proponent of the market approach, acknowledges that workers must have "choice in the selection of jobs" for a market in occupational safety and health to function. But he fails to acknowledge existing economic constraints on those choices. To Smith, poor people prefer high wages to safe jobs; they are not led to that choice by circumstance. Speaking of the market in health and safety, he suggests:

> The end result would be that people who do not value
> additional safety very highly relative to additional
> income—whether because they are poor, have high pain
> thresholds, or just do not care—would choose more wages
> over more safety. . . . Those who value additional safety highly
> would tend to take jobs in plants where safety is relatively
> cheap to ensure.[11]

Nevertheless, Congress was quite concerned with the distributional issues raised by occupational safety and health when it wrote the OSH Act. The American welfare state has tended to promote equal protection, and this legislation extended the notion to the workplace. Congress created a universal right to safe work. In doing so, it required the agency to achieve levels of protection that are not likely to correspond to what workers would choose in markets. Administrative rules were purposefully substituted for market processes with the understanding that the mandated level of risk reduction could depart significantly from what workers selected when they traded wages for protection.

Equally important, the OSH Act orders OSHA to try to equalize risk among workers. The obligation to equalize risk can be seen in several of the act's provisions. These have been noted previously but they are worth restating here. The act's expressed purpose is "to assure so far as possible *every* working man and woman in the Nation safe and healthful working conditions" (Sec. 2b; my emphasis). To this end, the act requires that each employer "furnish to each of his employees employment and a place of employment which are free from recognized hazards that are causing or are likely to cause death or serious physical harm to his employees" (Sec. 5a[1]). To enforce these rights, the secretary of labor is required, when issuing standards that deal with toxic materials or "harmful physical agents," to "set the standard which most adequately assures, to the extent feasible . . . that *no* employee will suffer material impairment of health or functional capacity even if such employee has regular exposure to the hazard dealt with by such standard for the period of his working life" (Sec. 6b[5]; my emphasis).

In short, Congress did not distinguish among hazards or attempt to regulate only particularly high-hazard industries. It sought to provide all workers with safe jobs. Smith's recommendations, then, violate the law's basic premises.

Private legal action is problematic on other grounds. Strict liability could replace negligence as the legal standard for damage suits. This change would liberalize the conditions under which workers could sue employers. Employers could be held responsible for system failures such as design flaws or random and uncontrollable breakdowns, as well as gross negligence. This should result in a large number of successful class-action suits over occupational diseases such as silicosis and hazards such as hearing loss. In the long run, these suits would provide incentives for employers to prevent hazards.

But costly settlements would threaten the economic viability of many industries. In the short and mid-term, they would probably bankrupt a large number of firms and a significant number of industries. Efforts to liberalize workers' compensation for silicosis and asbestosis have already resulted in government limitations on the recovery of medical benefits in some states.[12] Liability suits against the Manville Corporation (formerly Johns Manville) for asbestos exposure led that company to seek the protection of the bankruptcy courts.

It is probable that the proliferation of successful damage suits such as these would lead the federal government to cap employers' liability and create some form of public subsidy to affected firms. The Black Lung Program has already done this for the coal industry. The program subsidizes employers, restricts the ability of workers to sue them, and therefore limits the incentive effects of court action.

Nor are markets and court action apt to encourage worker participation in decisions over working conditions. As I stated earlier, collective action is necessary if worker participation mechanisms are to function effectively. But markets tend to discourage the kinds of collective actions that are necessary if workers are to take a more active part in determining the conditions of work. Markets are, after all, famous for encouraging privatism and individualism. As individual buyers of health and safety and sellers of labor, workers become self-interested utility maximizers. As rational individual actors, workers are most likely to respond to the market economistically, that is, by changing jobs or demanding higher wages rather than seek control over working conditions. Legal action is unlikely to correct this situation. Instead, it might encourage economism by focusing worker demands on compensation rather than prevention, and shift worker struggles from the shop floor to the courtroom.

The market-conservative approach, then, falls short on both of the relevant criteria. First, because workers are likely to undervalue health and safety, the market is unlikely to create incentives sufficient to produce a level of investment in health and safety appropriate to the hazards that workers face, or to be responsive to demands that workers would make under less constrained circumstances. Legal remedies are not likely to change this outcome substantially. Second, by discouraging collective action, this approach would make it difficult for workers to play a central role in plant-level decisions on working conditions.

Rationalizing Regulation

A second set of reforms seeks to maintain a leading role for public power but to make regulation more "rational" by introducing economic values into standard setting and enforcement. In this sense, it reflects the growing importance of neoliberal policy analysts and post-New Deal Democrats. To both groups, it is imperative that public officials take the impact of their decisions on capital investment into account when they select policy goals and that implementation strategies are chosen that minimize economic disruption caused by state action.[13] As we saw in Chapters 4 and 6, this view became increasingly popular after the recession of 1973–1975.

Applied to workplace safety and health regulation, three specific changes have been suggested: (1) the use of economic review to set priorities among hazards; (2) the adoption of performance rather than detailed design standards; and (3) the use of injury taxes in place of penalty-based inspections.

Making Regulation Efficient

Those who propose to rationalize regulation do not deny that worker protection raises equity issues, but they do suggest that worker rights represent one particular claim that must be balanced against the general claims of society. Since these general claims are premised on economic growth and capital investment, advocates of rationalizing regulation urge the adoption of one or more of the following methods to review standard setting and enforcement: cost-benefit analysis, cost-effectiveness analysis, and the regulatory budget.

Cost-benefit analysis has been proposed as both a decisional tool and a decisional rule. As a decisional tool, it requires that agencies identify, monetize, and quantify the consequences of regulation. The Carter administration promoted cost-benefit analysis in this form. As such, it was designed to give public officials a precise measure of the effects of proposed policies and one way of comparing various regulatory alternatives. It did not require agencies to reject actions when costs exceeded benefits. It did, however, force agency officials to justify these proposals. Adopted as a decisional rule, as the Reagan administration used it, cost-benefit analysis requires that policy-makers reject rules and policies whose net benefits do not exceed their net costs.[14]

Cost-effectiveness analysis can be applied to regulatory goals and methods. Applied to goals, it requires that agencies that are attempting to reduce several risks at once distribute the costs of regulation in such a way as to equalize the benefits gained across programs. For example, a dollar spent controlling cotton dust should be as effective in reducing the risk of byssinosis as a dollar spent in reducing the incidence of asbestosis. Applied in this fashion, cost effectiveness is similar in effect to cost-benefit analysis, as Zeckhauser and Nichols's summary of the case for a discretionary cost-benefit test indicates. Used as a decisional tool, cost-benefit analysis would, they claim,

> force OSHA to examine the consequences of its standards more
> closely . . . it would highlight inconsistencies in different
> areas; it might show, for example, that at current levels of
> stringency one standard costs $5 million at the margin per
> expected life saved, while another could be tightened at a cost
> of only $5,000 per expected life, thereby yielding 1,000
> times the OSH gain for its cost impositions. In such a case, by
> loosening the first standard and tightening the second, it
> would be possible both to increase longevity and to free
> resources for other uses.[15]

Alternatively, cost effectiveness has been used to recommend compliance techniques. The agency has been urged to select the most cost-efficient method to achieve whatever regulatory goal it chooses. Since neither approach requires that regulations pass a cost-benefit test, this methodology has been proposed as a compromise between those who support and those who oppose economic review. According to its proponents, this method leaves room for political and ethical considerations but also acknowledges the importance of costs to firms and the economy.[16]

The regulatory budget takes a different approach. Central authorities impose a limit—the budget—on the total costs to the economy that a single agency can impose each year. Within this budget constraint, regulators can choose what and how to regulate. This idea surfaced at the end of the Carter administration and was actively promoted by OMB. Because the Reagan administration dropped it in

favor of mandatory cost–benefit analyses, the idea remains unde-
veloped, and various issues remain unresolved. Such issues include
who would set agency budgets, how these figures would be determined, and whether agencies could "bank" regulatory costs from
year to year. Presumably Congress would set aggregate cost limits,
and the White House would review agency proposals to determine
whether they meet these legislatively determined budgets.[17]

Proposals to replace design with performance standards and to
substitute an injury tax for fines levied by inspectors are intended to
increase the employer's discretion in complying with health and
safety goals. Performance standards substitute health and safety targets, such as fewer lost-workday injuries or the reduction or elimina-
tion of worker exposure to a particular hazard, for detailed design
specifications. The firm then chooses among competing control technologies, ranging from personal protective devices to complex engineering controls. The agency, in turn, penalizes the firm when it fails
to meet its targets, instead of penalizing the firm when it fails to adopt
specific machinery or work practices.[18]

The injury tax is an economic incentive system that also maximizes
employer discretion in the choice of control methods. Its intellectual
roots can be traced back to economic arguments for effluent taxes to
control pollution and the concept behind experience-rating firms
under the workers' compensation system. Schultze's 1976 Godkin Lec-
tures on regulation stimulated interest in applying this approach to a
wide range of regulatory programs.[19]

Taking this approach, public officials would establish protective
goals by setting charges for injuries. These charges could be set at
various levels depending on how much protection was sought. If the
market model is taken as a standard, the charge could be set at the
level necessary to raise the costs of accidents to firms to the point
where they equaled the costs of prevention. Or charges could be set
above or below the level of economic efficiency to achieve more or
less than the economically "optimal" level of protection. In either
event, the injury tax would be cost efficient because employers would
be free to choose the least expensive method of reducing injuries. In
the case of workplace safety and health, an injury tax would allow a
firm to experiment with various remedies until it found the particular
mix that was suitable to its work environment.[20]

Rationalizers argue that these kinds of reforms are necessary to deal with the problems of resource scarcity and political account-ability raised by social regulation. Given resource scarcities, agencies accept the need to choose regulatory goals carefully and minimize compliance costs. Protective statutes, however, are often ambiguous; health and safety risks are widespread and numerous; and well-organized constituencies pressure agencies for the highest possible levels of protection. Thus, the agencies are likely to set regulatory goals that cannot be justified by efficiency or equity considerations.

In contrast, economic review helps agencies set priorities among competing goals and choose the most cost-efficient means of achieving them. It focuses attention on the implications of their choices and encourages them to make choices rationally. Cost–benefit tests, cost-effectiveness analyses, and the regulatory budget make agencies conscious of scarcity and force them to choose efficient control technologies.

The review process also helps hold agencies accountable to higher authorities and limits the debilitating effects of "special-interest" legislation on the economy. Because they supervise the process, White House officials enjoy a number of levers over agency policy, including oversight of the preparation of the analyses, the final determination of agency requests to regulate when costs outweigh benefits, and the power to set agency cost budgets. Presumably these officials would be in a better position to resist interest-group pressures for higher protection and bureaucratic pressures for expanded agency authority.

Corporatism Without Labor?

Whatever its promise, without major changes in the existing approach, economic review is apt to evolve into a distinctively American form of corporatism, one that encourages close contact between business and government but discourages participation by rank-and-file workers or organized labor. As a result, it is unlikely to be used to increase democratic decision making on protection or redistribute resources toward worker health and safety.

Economic review procedures are likely to limit rather than encourage investment in worker health and safety for several reasons. First, the monetization of costs and benefits usually devalues income and wealth transfers. All forms of economic review assess the consequences of regulation by adding and subtracting market prices. But market-determined prices result from exchanges made between unequal parties, and the review process valorizes these income and wealth differences. Any attempt to impute prices to untraded commodities—to use wage differentials to discover how workers value low- and high-risk work, or to discount a worker's future earnings to determine the present value of his or her life—does the same. Poorer workers are revealed to value their health and safety less highly than more affluent workers; old people, because they do not have future earning power, turn out to be worth less than young people.

Given marked economic inequality, the effects can be perverse. Cost–benefit analyses can recommend highly regressive projects and recommend against redistributional programs because of the relative economic positions of different income classes. One of the leading experts in this area suggests that this disregard for the distributional consequences of cost–benefit tests is likely to occur even when the analyst attempts to introduce distributional criteria into the analysis:

> Even if it were conceivably possible to secure permanent
> agreement . . . on the set of distributional weights to be
> attached to the benefits and losses of different income groups, it
> could not . . . be counted on to prevent the introduction of
> projects having markedly regressive distributional effects.
> Projects that would meet a weighted cost–benefit criterion
> could be such as to make the rich richer and the poor poorer if
> the beneficiaries were rich and many and the losers poor
> and few.[21]

A review effort that sought to calculate costs and benefits on the basis of individuals' market choices would have similar effects. As noted, several factors shape workers' tradeoffs between safer and better-paying jobs, including their income and wealth positions. A poor, unskilled worker is more apt to take a risky job than an affluent, professionally trained worker is. Since a worker's initial market position will shape how he or she values health and safety, standards based on this kind of benefit analysis will distribute protection according to these preexisting market inequalities.

Unfortunately, this point is often ignored; when it is acknowledged, it is usually sidestepped. In *Risk by Choice*, Viscusi admits that this distributional criticism is "fundamental" and "legitimate," but offers a weak response to it.[22]

> Efforts to promote present risk regulations on the basis that they enhance worker rights are certainly misguided. Uniform standards do not enlarge worker choices; they deprive workers of the opportunity to select the job most appropriate to their own risk preferences. The actual "rights" issue involved is whether those in upper income groups have a right to impose their job risk preferences on the poor.[23]

Instead of redistributing through regulation, Viscusi suggests accomplishing redistributional ends through direct resource transfers. Such transfers would give poor people the same opportunity that other workers have to turn down unsafe jobs.

This makes sense, in theory, as do arguments that resource transfers and the benefits of regulation could be calculated differently. Cost-benefit tests, for example, could be weighted to favor redistributional transfers by multiplying benefits to workers by some politically determined factor: the poorer the worker, the larger the number. But political realities argue against this outcome. Instead, review procedures are likely to continue along the established trajectory and take their benchmarks from the market, with its existing inequalities.

In addition, cost–benefit and cost-effectiveness tests usually result in the disaggregation of the universalist right to protection found in the OSH Act. Review procedures normally apply a neoclassical concept of efficiency to standard setting, and this is likely to produce decisions that are directly contrary to the act's mandate to equalize risk.

As discussed earlier, hazards are unevenly distributed across jobs and industries. Different individuals face different kinds of hazards at work. Moreover, some hazards are more difficult to control than others. Consequently, it is more expensive to protect some workers than others. Given the neoclassical economist's notion of efficiency, standards that provide equal protection to workers in diverse settings are irrational. They do not deliver the greatest net benefits, and they do not equalize the marginal costs of protection across jobs and industries. Efforts to use economic review to correct these problems— whether to set priorities or accept or reject particular standards—will result in the disaggregation of the universalist right to health and safety created by the act.

The CWPS intervention in rulemaking in the coke-oven emissions case provides an excellent example of how this has been done. When CWPS recommended that OSHA consider regulating risks in "other occupations with both higher relative risks and much larger absolute numbers" of workers at risk, it claimed to speak in the name of efficiency. Were CWPS actually arguing that OSHA should protect other workers, this recommendation might be less disturbing. But CWPS's critique of OSHA's proposed standard reflected the White House's desire to control the total costs of regulation. Moreover, it implicitly rejected the act's legislative mandate to reduce risks for *all* affected workers. In short, economic review resulted in recommendations that, if implemented, would have led to a lower level of investment in health and safety than that proposed by the agency or envisaged in the act.

The use of economic review to impose regulatory methods such as performance standards and injury taxes is also likely to limit investment in health and safety. Most occupational safety and health and safety professionals subscribe to the idea that there is a "hierarchy" of preferable controls beginning with control at the source, then proceeding to control of the transmission or dispersion of toxic substances, and finally to control at the point where the worker is directly exposed to hazards by personal protective equipment, work practices, and administrative controls. But there is a tradeoff between the costs of control technologies and their effectiveness, and the profit-maximizing employer will probably opt for the least costly rather than the most effective approach. Indeed, employers have resisted engineering controls because they are much more expensive than dealing with worker behavior.

Theory and experience suggest, however, that the hierarchy-of-control concept is valid. Control at the source minimizes the problem of worker error resulting from stress, fatigue, or boredom. In practice, PPDs rarely work as well as they are supposed to. Masks, for example, leak and rarely fit well; they make it difficult to breathe and impede communication among workers—in itself a hazardous condition. Workers often remove them or falsely believe that they are receiving adequate protection when they are not. Thus the employer's discretion to adopt this option should be carefully limited to maximize worker protection.[24]

Nonetheless, as we have seen, White House reviewers in every administration endorsed this view and argued for PPDs on cost-effectiveness grounds. They pressed OSHA to allow employers to use ear plugs to protect workers from noise, and masks to protect them from dusts and gases. Indeed, many of OSHA's critics argue for incentive mechanisms because they free the employer to adopt PPDs and work practices in the place of engineering controls.

Economic review is also likely to devalue worker participation. No administration has attempted to assign participation a positive monetary value in its own right. In the labeling controversy, OMB expressly denied the value of worker rights to know, apart from their specific impact on increased safety and health. In fact, economic review is apt to treat worker participation as a cost; labor time devoted to health and safety will weigh against protection in cost-benefit tests.

Recommendations to maximize the firm's discretion in the name of economic efficiency are also likely to devalue worker participation. In theory, an employer seeking to maximize the efficient production of health and safety might involve workers in an effort to take advantage of their knowledge and skills and increase their motivation to take care. But there are powerful economic incentives to employers to resist worker control. It is more probable that the rational employer will forgo the possible health and safety benefits of increased worker participation in favor of the known benefits derived from labor discipline and control.

In sum, proposals to rationalize regulation promise to increase the effectiveness of whatever resources are devoted to health and safety, but they are likely to discourage investment in worker health and safety and worker participation at work. In this way, they administratively rewrite the OSH Act and subvert the goals of an effective occupational safety and health program.

Labor-Oriented Approaches

Labor oriented approaches refer to workplace regulation in which workers and unions play a central role in policymaking through one or another variant of neocorporatism. These approaches are common in Western capitalist countries and constitute a distinct alternative to the liberal approach that dominates workplace regulation in the United States, as well as the two approaches to regulatory reform discussed above.

Not all labor-oriented approaches are radical. Like their American counterparts, many Western European labor movements have reached accords with business and accepted a subordinate place within a welfare capitalist system; production remains organized according to capitalist criteria. Even social democratic regimes presume a good deal of private control over investment and work.

Nonetheless, in many European societies, organized labor is more powerful in politics and economics than it is in the United States. As a result, workers play a greater role in policymaking—through social democratic and labor parties and union confederations at the national level—and exercise more influence over employers at the local level—through works councils and other forms of worker organization. Where labor is more powerful, occupational safety and health policymaking tends to reflect that power by providing unions and workers with institutional arrangements that facilitate participation in standard setting and enforcement. Some of these arrangements seem to increase investment in health and safety and worker control over the conditions of work.

Neocorporatism

Neocorporatist approaches to occupational safety and health combine tripartite standard setting with worker participation in enforcement. As in the United States, basic rights and responsibilities are codified in statutes, and the state has the power to inspect and fine firms for violating standards. But this approach encourages business and labor to use government as a forum to negotiate general policy and particular standards. It also deemphasizes penalty-based inspections. At the same time, neocorporatist approaches encourage worker participation at the plant level by creating statutory health and safety committees and occupational health programs. Thus, neocorporatism combines bargaining over policy and standards by the leaders of business and labor—a highly centralized form of decision making—with decentralized systems of local enforcement.

Advocates of tripartite standard setting claim two major advantages for it. First, tripartitism is supposed to reduce conflict and assure that employers comply with standards. Because policymaking boards and commissions provide union organizations and business confederations with regular, guaranteed access to government decisionmakers, tripartitism encourages negotiations that lead to mutually acceptable agreements. Because employers are given formal representation, they are more apt to implement the agreements that are reached. And, to the extent that the officially sanctioned employer associations speak for business as a whole, are compulsory membership organizations, and provide essential services to their constituents, they are likely to be able to force recalcitrant firms to comply with negotiated agreements.[25]

For several reasons, neocorporatism also is supposed to maximize the flexibility of policy. For example, secondary issues that arise within the general parameters established by legislation can be resolved through regular negotiations between labor and business. Because these negotiations are informal—they do not follow the judicial model of conflict resolution characteristic of the American system —participants can reach compromises that balance their interests across a wide range of related issues. This system encourages flexibility by facilitating package deals in which concessions are made on some points in return for victories on others. Finally, because these negotiations can be organized to design and implement long-term plans, organized labor has more certainty that bargains struck today will be enforced tomorrow. Unions are therefore more likely to trade off present gains for future benefits, thereby allowing employers the opportunity to come into compliance more slowly and at less cost.

The following comparison between the logic of the liberal and neocorporatist approaches illustrates the point: The American approach relies on an expansive, statutory commitment to worker health to be implemented by expert administrators. It does not rely on tripartite arrangements to decide policy or set standards. In fact, unions have resisted negotiating standards or compliance agreements because they see them as a departure from the statutory rights granted to workers in the OSH Act. Lacking institutional mechanisms to assure that concessions in the present will lead to implementation in the future, the labor movement expects that any departure from the act's ethic of protection will lead them down a slippery slope of deeper and deeper cuts without any corresponding benefits.

Given the liberal approach, and the unions' own political and economic weakness, organized labor is acting rationally when it adopts this strategy. Unless forced to, employers are unlikely to compensate reasonable workers with future benefits. Under these circumstances, each compromise and every concession is a management victory rather than a step toward a long-term goal of hazard reduction. Moreover, once they take strong positions, unions are led to defend them in order to maintain their reputation as powerful political actors and their influence over public officials and their own members.

In contrast, a neocorporatist approach to standard setting allows the state to commit public authority to an expansive definition of protection but, within this general frame, forces unions and employers to negotiate priorities and timetables and implement these settlements. For example, organized labor and employer groups can negotiate the pace of standard setting and the application of standards to particular firms and industries in keeping with a plan that seeks to reduce risk across all occupations in the long run. Organized labor can make other concessions with some certainty that these ultimately will be rewarded. Given an enforceable agreement to raise investment in health and safety, cost-effectiveness and cost-benefit tests can then be used to channel new investment to the most immediately productive uses rather than question the logic of equal protection. Organized labor also can negotiate review procedures that place a positive value on worker participation.

Indeed, it might prove easier to gain worker acceptance for incentive mechanisms and employer experimentation if workers believed that employer compliance could be easily monitored and failed experiments corrected. Requests for variances from specified design changes, approvals to experiment with new control technologies, and special dispensations to economically distressed enterprises could be negotiated industry by industry, or firm by firm, in return for commitments to detailed plans to increase investment when it became economically feasible. Where such investment might never be economically feasible, workers could negotiate concessions on other issues of concern to the affected workers, such as job security or participation in firm governance. Ideally, the entire social regulatory enterprise would take place within the context of economic planning designed to facilitate democratic determination of which distressed industries are worth saving, and at what cost in accidents and injuries.

Under these conditions, the idea of universal protection would be treated as a goal rather than an immediately enforceable rule. It is not unimaginable that organized labor might endorse a review program based on this premise. After all, it is neither impossible nor unreasonable to rank the various threats to worker health and safety and take up the most pressing ones first. An index based on what is known about a substance's morbidity and the degree of exposure to it is one obvious measure. But, in the United States, health and safety reformers have rejected this concession because they think it might lead to the protection of some workers at the expense of others. It has. Under different arrangements, however, economic review might serve different interests.

Advocates of the neocorporatist approach to workplace safety and health also argue that it increases worker participation in enforcement. As a rule, these systems do include stronger statutory commitments to worker participation at the workplace, including legally required health and safety committees, safety representatives, works councils, and rights to know about hazards and refuse hazardous work. The existence of these rights and institutions, in turn, facilitates worker activity at the enterprise-level because workers know about, and recognize, hazards and have the organizational capacity to pressure employers to reduce them.

Antidemocratic Tendencies in Neocorporatism

The historical record indicates that corporatism can easily devolve into an elitist form of policymaking that discourages democratic participation by the rank and file and advantages only a few workers.

Informal negotiations between organized labor and business groups can lead to a kind of administrative politics in which deals are struck that serve only the leaders or the best-organized interests. As a rule, the degree to which neocorporatism actually advantages workers depends on how much power workers really do exercise under these arrangements.

In regard to workplace regulation, four rather stringent conditions are necessary if neocorporatism is to lead to higher levels of investment in occupational safety and health and facilitate effective worker participation in plant governance. First, and probably most important, the labor movement must be strong enough to represent workers' interests effectively in negotiations with employers and force elected officials to implement these negotiated agreements. Second, the labor movement must make occupational safety and health a priority; otherwise, it will not devote its resources to the issue, and worker interest will wane. Third, workers and worker representatives must be able to function independently of management. This means that they must enjoy statutory rights to have formal access to state inspectors, to refuse hazardous work, to know about hazards, and to veto or withhold consent to management decisions concerning health and safety. Finally, workers must have access to the resources necessary to police the workplace, monitor firm compliance with relevant standards, and participate in enforcement, including information, organizational capacity, time, and money.[26]

The comparative experience suggests that these preconditions are rarely met in full because of deficiencies in legislation and less-than-aggressive trade union strategies. In Britain, for example, joint health and safety committees or, in small firms, safety representatives have been mandatory since the mid 1970s. A 1974 law requires that occupational safety and health representatives be appointed, and a 1975 statute gives the unions the sole right to choose these representatives. Subsequent amendments have given health and safety representatives rights to inspect enterprises, to be paid for health and safety training, and to investigate accidents.[27]

These rights are limited in several ways. The committees' powers are qualified, and employers can veto decisions that require resource expenditures or changes in personnel practices. Moreover, worker rights rest exclusively on trade union action; employees do not have rights as individuals. The unions select worker representatives to the joint health and safety committees, and these union-selected representatives exercise the rights to participate in inspections and refuse hazardous work.[28]

Compounding the problem, the British labor movement has not emphasized health and safety. Consequently, the health and safety committees and safety representatives do not play an active role in the determination of working conditions. Worker participation has not resulted in control over plant-level decisions that affect health and safety, and workers have not sustained interest in these programs.[29]

West Germany reveals a similar pattern of limited rights and trade union strategies. There, works councils and health and safety committees enjoy broad statutory powers to participate in general plant governance, including the right to veto health and safety decisions. The councils, the committees, and the rights are mandated by law; workers participate in occupational health and safety programs through them.[30]

But worker rights are limited in several ways. Members of works councils are a minority on the safety committees and have few operative powers. As in Great Britain, the rights of individual workers are restricted. The right to refuse dangerous work is sharply limited, and workers cannot call in safety inspectors. Moreover, worker rights rest almost entirely on union initiative for their force, and the union movement has not made health and safety a priority issue. As a result, the rights that workers enjoy are largely unexercised. In contrast, business groups enjoy privileged access to standard-setting bodies, and management-oriented public health professionals dominate the system. Thus, despite their statutory rights, workers do not participate in most of the important decisions concerning occupational hazards, including the application of new technologies.[31]

The Social Democratic Variant

The social democratic approach is a variant of neocorporatism distinguished largely by the power that organized labor enjoys in politics and reforms designed to facilitate worker control over plant-level decisions. Thus, this variant combines more effective union participation in national standard setting with more active worker participation in enforcement. In both instances, strong legislation is reinforced by more radical labor strategies. Standard setting, for example, includes organized labor on an equal footing with business. Health and safety committees exercise real power over working conditions. Individual workers enjoy full rights to participate in inspections, to know about hazards, and to refuse hazardous work.

Advocates of this approach suggest that these arrangements provide the benefits of neocorporatism but avoid the problems that often surface with the neocorporatist approach. Most important, greater worker participation in this approach limits the antidemocratic tendencies of neocorporatism in several ways. By involving workers in plant programs, the social democratic approach raises their consciousness about health and safety and increases the likelihood that they will take an active part in these programs. Workers also are provided with organizational forums that help them coordinate their activities and lower the costs of participation. Additionally, hands-on experience in meeting plant-level health and safety goals provides workers with the information to assess central decisions about what is practical and necessary. Worker participation also encourages workers to pressure unions to represent their interests in health and safety. Unions are then more likely to pressure the state to take health and safety seriously, and to pressure employers to comply with government regulations. Finally, worker participation reduces the burden that occupational safety and health regulation places on public officials. Properly trained, organized, and represented, rank-and-file workers can observe firm practices, including compliance with standards and negotiated agreements, and pressure plant supervisors more effectively than a distant national agency can. In fact, a mobilized workforce should reduce the need for government inspectors to monitor working conditions.

Like market-conservative approaches, the social democratic approach also builds on private action in two ways. First, workers play a greater role in enforcement than with any other approach. Second, with better information and stronger rights, workers are able to make informed choices about the jobs they wish to take or, if they are interested in taking risky jobs for higher wages, to make informed decisions about that tradeoff. But this private action occurs within a different institutional setting. Most important, political mechanisms exist through which workers can democratically determine their rights and protective goals. These institutions, in turn, help to discourage privatism and individualism and encourage workers and unions to make the organization of work a political issue.

The social democratic approach should also overcome some of the problems encountered when workers rely exclusively on collective bargaining to reduce occupational hazards. Collective bargaining is a relatively ineffective way to change the conditions of work. Union contracts rarely cover all workers or provide covered workers with equally strong protections. Instead, they result in a highly variegated pattern of worker rights. This is apt to encourage divisions among workers and decrease the likelihood that the labor movement will make health and safety a priority. Even when contractual agreements specify worker rights, these rights are not apt to be enforced without worker representation in statutory health and safety committees or works councils.[32]

In sum, social democratic arrangements have the potential to provide workers with greater information about, and leverage over, working conditions than other approaches. In-plant organizations can monitor the hazards faced by employees and make sure that state inspectors do their jobs. By applying independent pressure on employers, they can help state agencies assure that firms remedy unsafe conditions. Thus, if workers participate, a social democratic approach can use the state to supplement, reinforce, and build on the virtues of private action without succumbing to the limits inherent in market-based approaches.

Sweden

Only the Scandinavian experiments in social democracy actually approximate the conditions outlined above. Sweden in particular illustrates the potential impact of social democracy on the regulation of working conditions. There, the majority of workers are organized and the Social Democratic party has governed for all but six years since the Great Depression. The state is more interventionist than is normal in capitalist democracies and enjoys a wide variety of powers over capital flows and labor markets. Occupational safety and health policy is equally well developed.

Participation in workplace decisions in Sweden is justified in political and philosophical terms. The 1977 Work Environment Act declares its aim to be "for work to be arranged in such a way that the employee himself can influence his work situation." The Swedish Labor Ministry has interpreted "the underlying idea" of the act to be that "active participation by employees can establish a form of working life characterized by partnership and shared responsibility, security and meaningful jobs and job satisfaction."[33]

Standard setting is tripartite in form. The Worker Protection Board (Arbetarskyddsverket, or ASV), the Swedish counterpart to OSHA, develops rules in consultation with tripartite committees. The committees include representatives of the Landsorganisationen (LO) and the Tjänstemannens Centralorganisationen (TCO), the Swedish central blue-collar and white-collar labor organizations, and the Svenska Arbetsgivareföreningen (SAF), the employer organization responsible for negotiating economywide collective bargaining agreements. Committee meetings are informal and coordinated by ASV. The management and labor groups serve as interlocutors between ASV and individual firms and union locals. They circulate standards proposals to their members and communicate members' proposed revisions to ASV. The agency's powers are substantial: it does not have to keep elaborate written records of its meetings or provide for formal hearings and comments; it enjoys final authority to set standards; legal challenges to standards are not permitted.[34]

Although Swedish employers have input into general occupational safety and health policy and ASV decisions, the strength of the Swedish labor movement and the state's leverage over the process of capitalist investment have encouraged public officials to adopt strong health and safety programs, including extensive worker rights to participate in plant-level decisions. The 1976 Joint Regulation of Working Life Act and the 1977 Work Environment Act give workers the right to participate generally in firm governance and specifically in the determination of working conditions. Workers are guaranteed a variety of statutory rights, including the right to refuse hazardous work, to stop work in imminent-danger situations, to know about hazards, to participate in plant activities, and to supervise the operation of in-plant health services. Employers are required to fund factory-based medical clinics for occupational safety and health. Health and safety committees supervise these in-plant programs, and shop stewards are authorized to monitor and investigate workplace hazards. They also enjoy the right to shut down plant operations temporarily in the face of imminent hazards.[35]

Collective bargaining between SAF and the two central labor organizations has reinforced these statutory rights. In 1979 employers and employees agreed to increase the autonomous powers of the health and safety committees to veto the appointment of health and safety experts, to organize the medical service, and to give worker representatives a permanent majority on the health and safety committee.

Reflecting organized labor's political power, ASV standards are strict—most of them are stronger than the equivalent OSHA standards. In 1978, ASV exposure levels were lower than OSHA exposure levels in over half of the 145 cases in which both agencies had standards. In contrast, OSHA levels were stricter than ASV standards in less than 3% of the cases.[36]

The situation in Sweden also indicates that when organized labor negotiates from a position of strength, standards are likely to be decentralized and more easily revised. Labor unions frequently agree to compliance schedules that take the particular economic conditions of firms into account. In turn, SAF attempts to secure the compliance of individual firms with government standards. The ASV standards also are more frequently updated than OSHA standards and more apt to be based on current research. Finally, employers are much less likely to challenge ASV rules.[37]

By no means has Swedish social democracy eliminated the tendency for capitalist production to undersupply health and safety. But more extensive state control over investment and labor markets and more politically conscious and well-organized labor movements create powerful countervailing forces that have led to increased investment in health and safety—as ASV's stricter standards suggest —and increased worker participation in decisions about working conditions.

The Contradictions of Social Democracy

As the analysis above indicates, the success of labor-oriented approaches varies with the strength of the labor movement and the
willingness of public officials to act against the interests of private
investors and firms. Workers must be well organized and take occupational safety and health seriously. Public officials must be held
accountable on this issue. Herein lies the dilemma for those who
advocate a social democratic approach. Since social democratic
regimes govern capitalist democracies, private property in production, profit-seeking investment, and labor markets continue to operate.
In combination, they encourage firms to disregard hazards and employees to focus their efforts on short-term material gain. Although the
political organizations of workers and reformist administrations can
counteract these tendencies, the normal operation of the system undermines reform. Thus the state and labor must struggle continually
to maintain the ground that they have won.

If the union movement shifts its attention to other issues, the standards set through tripartite negotiations are apt to weaken. The European cases discussed above suggest how labor strategies affect
occupational safety and health policy. Both Great Britain and Germany adopted worker participation programs in the 1970s, but in
both countries, worker rights far outstrip the reality. Neither the British
nor the West German labor movement has responded to the rights
that they enjoy. The British program gives unions the rights to select
safety representatives and take part in public programs, but the
Trades Union Congress (TUC) has not taken advantage of these opportunities. Few unions employ full-time health experts to supplement
existing safety representatives. The TUC does not conduct an independent research effort, as LO does in Sweden.[38] Similarly, the West
German program depends directly on union participation, and the
unions have proved indifferent; the works councils have not focused
on the issue. Worker participation is limited and, as in other areas of
codetermination, employers continue to make most of the important
decisions about working conditions.[39]

Social democratic regimes also suffer from a kind of political-
economic cycle that undermines reform. Brought to power by a
mobilized working class, left-wing parties are able to build stronger

states and adopt more redistributive welfare and economic policies. But, once in power, social democratic parties become responsible for managing a capitalist economy. The need to reconcile social policy with capital investment and international competition emerges as it does in more liberal regimes. Social democratic parties are held responsible for economic crisis, and if economic problems become severe, they may moderate their programs.[40]

Nonetheless, social democratic regimes stand a better chance of prevailing in the face of these tendencies than do liberal regimes because of the existence of more radical worker organizations and greater state control over investment. To the extent that workers remain mobilized and committed to increasing public control over investment, the impact of economic decline on social policy can be contained. Most important, occupational safety and health need not suffer to the extent that it does in liberal-pluralist societies.

The Liberal Approach in Comparative Perspective

The preceding analysis brings us back to the point raised at the outset. Neither market-conservative nor neoliberal approaches to regulatory reform are likely to solve the problems inherent in the liberal approach to workplace regulation because neither successfully confronts how the political and economic power of business discourages investment in health and safety and worker participation in the determination of working conditions. Nor does either provide a compelling alternative to the statism of the liberal approach. Market-conservative proposals to substitute private action for state power rest on an unrealistic theory of the market that ignores how class inequalities shape the distribution of risk and protection and frustrate efforts by workers to protect themselves from hazards. Neoliberal proposals to rationalize regulation synthesize the values of capitalist production and regulation, but in a way that almost invariably argues for less, rather than more, protection and devalues worker participation—either because economic theory demands it or because public officials, dependent on capitalist investment, are reluctant to challenge these assumptions.

The discussion of labor-oriented approaches to workplace safety and health helps to clarify these issues by suggesting alternative ways of combining public and private action in capitalist democracies. The social democratic approach in particular argues for a radically different approach to the relationships among capitalist production, worker participation, and state intervention. State action services and is subordinated to the political and economic organizations of workers. State-enforced rights facilitate rather than replace collective action by workers. Thus, the social democratic approach promises to make workplace safety and health policy simultaneously more effective and more democratic.

The social democratic approach is not about to be adopted in the United States in the near or mid-term. Nonetheless, lessons can be learned from it. First, it underscores the narrowness of the American debate about regulatory reform. In addition, it provides several concrete suggestions about how reformers might approach social problems of this kind in the future. I now turn to this second set of issues by considering the future of social regulation in the United States. At the same time, I return to the theoretical issues raised at the outset of the book.

[9]
Conclusion

This book has developed along three related tracks. On the most general level, I have sought to contribute to a theory of social reform by analyzing the contingent character of business power and the possibilities and limits of anti-capitalist state action in the United States. At the same time, I have drawn on a framework derived from that theory to explain OSHA's inability to protect workers from occupational hazards. Finally, the theory and framework have been used to consider proposals to restructure American occupational safety and health policy. Here I bring together the conclusions reached along each of these tracks and weigh their implications for the future of social reform in the United States.

At the outset, I argued that the structure of capitalist democracy creates powerful obstacles to anticapitalist reform. Nonetheless, theory and history suggest that the ends served by the state are variable; the state can act in the interests of workers. But if reforms are to serve workers, two things have to be accomplished. First, reformers must increase the ability of public officials to take actions counter to business interests by loosening the state's dependence on private capital investment. Second, reforms must actively involve workers in the implementation of social policy. These two preconditions are also mutually reinforcing. When workers are actively involved in policy-making and implementation, they are not likely to abandon reformist public officials at the first sign of economic decline; public officials who are free to pursue worker interests may be less sensitive to the problem of business confidence; and workers actively involved in reform are less likely to embrace economism as a political strategy.

The rise and fall of OSHA illustrates both faces of social reform in capitalist democracies. On the one hand, it makes clear that the capitalist state is capable of passing social legislation over the opposition of business and in the interests of workers. With the passage of the OSH Act, the mobilization of new social and political movements, the White House's interest in them, and the determined efforts of labor activists led to a major extension and redefinition of worker rights. On the other hand, the failure of workplace regulation makes clear how business mobilization can combine with the structure of capitalist democracy to frustrate the implementation of anticapitalist reforms. In a period of economic uncertainty, political opposition by business was translated into a concerted and ultimately successful attack by the White House on OSHA.

The liberal approach to regulation left the agency particularly vulnerable to these political and structural forces: an overly statist but insufficiently radical approach to implementation failed to confront the obstacles to successful reform. Because it remained limited vis-à-vis the larger processes of production and capital investment, government remained highly vulnerable to the problem of business confidence. Moreover, because command-and-control regulation did not actively involve workers in the reconstruction of the workplace, the OSH Act failed to create the social and political support that might have helped public officials sustain reform in the face of business mobilization and economic decline. In this sense, the OSH Act proved counterproductive; it focused political opposition on the agency without creating the political and institutional preconditions for effective state action.

Worker Agency and the Social-Reform Cycle

In contrast to other studies, the present book suggests that worker agency—participation at the workplace and in politics—is a critical component to successful occupational safety and health regulation. Rank-and-file mobilization and the development of a movement for workplace reform created the preconditions for the passage of the OSH Act in 1970; subsequently, worker demobilization on the shop floor and in Washington undermined the implementation of the act. This study also suggests that regulatory reform must build on, and promote, worker agency if the OSH Act is to be implemented.

This problem of worker agency is doubly disturbing. On the one hand, it leaves public policy subordinated to a particularly vicious social-reform cycle in which the state's ability to serve workers' interests changes with movements in the economy.[1] Although the state can and does respond to popular pressure, social reform is vulnerable to economic cycles and countermobilization by business. Poorly organized, neither the poor nor the middle class can resist this pressure. Better organized, the labor movement has chosen a political strategy that leaves it overly dependent on the goodwill of public officials. Fearful of the impact of public policy on investment, labor leaders often abandon reforms when business confidence sags. This is especially true when social policies seek to regulate business in the interests of workers *as a class*. When that happens, countermobilization by business is apt to be intense, and social reform is likely to prove particularly vulnerable to rationalization.

Nowhere is this dynamic clearer than in the history of postwar social policy. The Great Society reformers did radical things; social regulation challenged the market and asserted a new, positive conception of human interests beyond the values of production. It created state-enforced social rights for people who could not protect themselves—consumers, neighborhood groups, and workers. Some of the

various health, safety, and environmental laws even encouraged
people to participate in government. Workers can accompany OSHA
inspectors; environmentalists can sue the EPA to implement clean air
and water statutes. But, at bottom, reformers created a paper state—
long on statutes and regulations, but short on effective social power—
because the state was left vulnerable to the punishing effects of the
market and the debilitating consequences of the political powerless-
ness of the people it was designed to serve.

Some liberal reformers continue to defend the Great Society and
explain its failures in conventional political terms. The climate of
opinion changed in America, they maintain, and support for reform
eroded. In another time, when better economic and political conditions prevail, or the underclass mobilizes, Americans will once again
accept their social responsibilities to the poor and the disadvantaged;
the state will again embrace reform. Then, with the resources and
political support they need, public officials will implement the statutory commitments of the Great Society and press toward even more
radical forms of liberalism.[2]

In one respect, this political argument is compelling. In the case
of OSHA, business opposition and the decline of organized labor
powerfully limited what government could do to protect workers from
occupational hazards. In another respect, however, this account of the
demise of the Great Society is mistaken. As public opinion polls indicate, workers have not abandoned their commitments to social reform. In fact, majorities continue to support tax programs that reduce
economic inequality; regulatory programs that control air and water
pollution, regulate consumer product safety, and protect workers
from occupational hazards; and public employment projects that
create jobs.[3]

Thus this political explanation for the demise of the Great Society
begs what is perhaps the most important question. Why have liberal
reforms been so vulnerable to changes in the economic environment
and the political opposition of business? It is by no means obvious that
laissez-faire ideas should resonate so loudly in the corridors of government or that economic decline should lead so directly and effectively
to the retrenchment of the welfare state. To the contrary, the inability
of a capitalist economy to provide workers with a secure and steadily
advancing standard of living in the 1970s and early 1980s, in com-

bination with a political attack on hard-won social reforms, might have led to a classwide mobilization in defense of liberalism. It did not.

The Future of Reform

If the analysis presented here is correct, it suggests two steps toward rethinking reform. First, social reformers, especially radicals who seek to loosen systemic obstacles to anticapitalist reforms, must begin with an analysis of the welfare state that takes explicit account of its structural limits and, simultaneously, the conditions under which they can be loosened. Second, the relationship between private and public action must be reconsidered and policy proposals fashioned that take maximum advantage of the opportunities for social change in liberal democratic systems.

The constraints on reform have already been discussed, and they point toward new reform strategies. The account of policy failure suggests that if reforms are to be sustained, they depend on independent political *and* economic activity by the people most directly served by reform. Most important, reformers must rethink the relationship between private and public action, and the role that reform can serve in facilitating both.

As a rule, Americans are trapped in a dichotomous way of thinking about the political economy that counterposes the state and the market and fails to consider the complex relationships between them. As a result, Americans swing back and forth between bouts of enthusiasm for state intervention and moods of deep distrust of all forms of public life—leaning first to "big government" and then to the "free market." It is commonplace to point out that neither image is accurate. In comparative perspective, the American government is not very big; in historical perspective, the market is surely not free. There is another, more important point to be made about this hyperbolic and distorted view of the political economy: it obscures the problems and the possibilities of acting in either sphere. Reacting against first the state and then the market, Americans fail to develop either to the point where it could serve democratic purposes. At least domestically, the state remains limited. At the same time, effective private action is fundamentally constrained by the obstacles thrown up to coordination and collective action by workers and consumers in capitalist markets.

Thus, taking reform seriously and seeking to empower people within this system requires a strategy that deals with both public and private spheres. The obstacles to effective private action must be lowered and collective action facilitated. Concurrently, the state must be reorganized so that it too can act effectively. Both spheres must be joined by a common ethic of participation.

As we have seen, the argument for private action is usually made by conservatives, who mean to confine social practices to the pursuit of private interests in markets; they question the efficacy and legitimacy of almost all forms of state intervention. Reformers need not adopt this view to take the point that liberal reforms have leaned too heavily on the state and that state action has been poorly thought out and poorly organized. As I have tried to demonstrate in this book, there are costs to statism, and benefits to be derived from close attention to which forms of state intervention help workers use public authority to promote self-organization and self-protection.

The issue of occupational safety and health suggests some specific illustrations. As a general rule, state action seems most effective as a way of determining a national, comprehensive approach to the problem of health and safety. Within that framework, there is a good deal of room for local efforts. Rather than rely on factory legislation, government can create rights to participate in the determination of working conditions and mandate in-plant mechanisms that involve workers in plant-level decision making. Public policy can then be used to coordinate central decisions about resource allocation and long-term health and safety goals with the local activities.

The discussion of the social democratic approach suggested one possible framework, but, for the moment at least, it can serve only as a frame of reference. Few Americans demand this approach to reform, and if it is imposed from above, it will turn out to be more corporatist than democratic. In any event, social democratic policy approaches do not guarantee that workers will organize, that they will define their interests broadly, or that they will take advantage of opportunities to participate in the implementation of protective policies.

Nevertheless, the range of alternatives considered points out some things that are ignored in American debates. First, it directly raises the question of how to recombine production, politics, and markets to maximize both participation and effective state action. Second, it suggests that organized labor and middle-class reformers must reorient their approaches to reform. Both must replace the liberal form of state intervention with one that is simultaneously more participatory and authoritative. Both must increase public control over the levers of investment. The latter is particularly important because the relationship between public authority and private investment establishes one of the fundamental parameters of individual and group decision making, including the decision by workers to seek short-term material gain or long-term political changes, and the decision by public officials to challenge private property rights or rationalize social policy.

Reform movements must also look for issues that simultaneously raise demands for participation and effective state action. Only then can they escape the private–public dualism that imprisons American political thought. Demand for a strong occupational safety and health program, built on mandatory enterprise-level programs and coordinated through a national health plan, is one example. Environmental protection organized around community boards and regional assemblies is another. In both instances, demands for protection from the hazards of capitalist production are combined with demands for increased participation in the organization of work and greater public control over investment.

Moreover, these issues appeal across occupational, income, gender, and racial lines. The demand for workplace safety and health, for example, can be formulated as a general claim for a democratic approach to basic decisions about the allocation of resources to competing uses, such as health care or hotel building. Of course people demand health care today, but their demands are rarely raised in this broader context—tied to demands for decision-making power and institutions in which that power might be exercised. Instead, they lobby for larger budgets and stronger programs. This conventional approach encourages divisions based on particular claims: for consumer, worker, or environmental protection; health care for the aged; housing for the poor. By focusing on the process by which allocational decisions are made, the common interest in securing greater public control over the levels and kinds of capital investment in society is underscored.

Both organized labor and middle-class reformers also must change their political strategies. The labor movement must seek more than social rights and adequate budgets. It must demand increased worker participation at work as it seeks greater leverage over the political system. It must seize on issues that cut across traditional industrial and sectoral lines and help to unite diverse constituencies. Concurrently, middle-class reformers need to extend their vision of participatory democracy to the workplace and the organization of production. They too must recognize the importance of public control over investment and worker participation at work for democratic politics.

One final point remains. This book began with a theoretical account of the structural obstacles to successful social reform in capitalist democracies, and the case of OSHA illustrates how difficult it is to use state action to serve workers. Nevertheless, I have recommended proposals that are self-consciously reformist, which might appear to contradict the basic thrust of the preceding analysis. Indeed, it is possible to find in the OSHA story compelling reasons to abandon reformism entirely.

There is another lesson here, however. By leaving the debate over the future of reform and social policy in the United States to conservatives and neoliberals, radicals abandon whatever ground can be gained by challenging the reigning orthodoxy with concrete proposals that move society in a different direction. To gain this ground, however, reformers must develop a subtle and nuanced understanding of public policy that combines a radical vision of what another society might accomplish with a close look at how state action contributes to, or undermines, progress toward that goal.

Notes

Introduction

1. See the public opinion polls cited in Chapter 5.
2. Public Law 91-596, Sec. 2b (my emphasis).
3. Ibid., Sec. 5a(1).
4. Ibid., Sec. 6b(5) (my emphasis).
5. Conversations with Joel Rogers were enormously helpful in clarifying these issues. For a penetrating account of American labor law, see his "Divide and Conquer: The Legal Foundations of Postwar U.S. Labor Policy" (Ph.D. dissertation, Princeton University, June 1984).
6. This view is popular among economists and policy analysts influenced by the neoclassical economics paradigm. See, for example, James C. Miller III and Bruce Yandle, *Benefit-Cost Analyses of Social Regulation* (Washington, D.C.: American Enterprise Institute, 1979), chap. 1.
7. Charles L. Schultze, *The Public Use of the Private Interest* (Washington, D.C.: Brookings Institution, 1977); Richard Zeckhauser and Albert Nichols, *The Occupational Safety and Health Administration—An Overview*, pp. 103–248 in *Study on Federal Regulation*, Vol. 6, S. Rept. 13, 95th Cong. 1st sess., 1978; Robert S. Smith, "Protecting Workers' Health and Safety," in Robert W. Poole, Jr., ed., *Instead of Regulation* (Lexington, Mass.: Lexington Books, 1982), pp. 311–338; Eugene Bardach and Robert A. Kagan, *Going By the Book: The Problem of Regulatory Unreasonableness* (Philadelphia: Temple University Press, 1982), a Twentieth Century Fund Report.
8. The most compelling theory of how group conflict shapes regulatory policy is to be found in James Q. Wilson, "The Politics of Regulation," in James W. McKie, ed., *Social Responsibility and the Business Predicament* (Washington, D.C.: Brookings Institution, 1974), pp. 135–168. Although there is some ambiguity in how his framework should be applied to occupational safety and health policy—it can be viewed as a case of either "interest-group" or "entrepreneurial" politics—Wilson's account provides the analytic foundations for an interest-group explanation of OSHA. For efforts to chart the impact of political conflict on social regulation in general and OSHA in particular, see David Vogel, "The 'New' Social Regulation in Historical and Comparative Perspective," in Thomas K. McCraw, ed., *Regulation in Perspective* (Cambridge: Harvard University Press, 1981), pp. 155–185; Mark Green and Norman Waitzman, *Business War on the Law* (n.p., Corporate Account-

ability Research Group, 1979); David P. McCaffrey, *OSHA and the Politics of Health Regulation* (New York: Plenum Press, 1982); Graham K. Wilson, *The Politics of Safety and Health* (New York: Oxford University Press, 1985).

9. There is a growing consensus in the literature on social regulation that different political-institutional arrangements play an important part in shaping national regulatory decision making by structuring the environments in which competing groups lobby for policies. See, for example, Graham K. Wilson, *The Politics of Safety and Health*; Ronald Brickman, Sheila Jasanoff, and Thomas Ilgen, *Controlling Chemicals* (Ithaca, N.Y.: Cornell University Press, 1985); Joseph L. Badaracco, Jr., *Loading the Dice* (Boston: Harvard Business School Press, 1985). These works were published after this book went into production, and I was not able to engage the authors' specific arguments in the text. However, I do wish to note several things. First, this work can be read as a contribution to what has been called the "new institutionalism," because in considering the relationship between economic structure and political-reform strategies it focuses on how public and private institutions affect the ability of workers and public officials to secure the implementation of social reforms. Nonetheless, this work differs from many of the new-institutionalist accounts in that I do not center the analysis on the organization of government itself, but my emphasis is not incompatible with those accounts. To the contrary, a complete theory of social regulation must consider all three factors: economic structure, political strategy, and governmental organization.

10. The literature on the capitalist state is large and growing. For two useful summaries, see Bob Jessop, *The Capitalist State* (New York: New York University Press, 1982); Martin Carnoy, *The State and Political Theory* (Princeton: Princeton University Press, 1984).

11. Karl Marx, *Capital*, vol. 1 (New York: Modern Library, n.d.), pp. 514–553.

12. V. I. Lenin, "The State and Revolution," in Lenin, *Selected Works* (London: Lawrence and Wishart, 1969), p. 273.

13. On the problem of business confidence, see Claus Offe, *Contradictions of the Welfare State*, ed. John Keane (Cambridge: MIT Press, 1984), pp. 119–129; Charles E. Lindblom, *Politics and Markets* (New York: Basic Books, 1977), chap. 13; Fred Block, "The Ruling Class Does Not Rule: Notes on the Marxist Theory of the State," *Socialist Revolution* 7, no. 3 (1977): 6–18. Michal Kalecki was the first to identify this phenomenon. See his *The Last Phase in the Development of Capitalism* (New York: Monthly Review Press, 1972), pp. 75–85.

14. See Adam Przeworski, "Material Bases of Consent: Economics and Politics in a Hegemonic System," in Maurice Zeitlin, ed., *Political Power and Social Theory*, vol. 1 (Greenwich, Conn.: JAI Press, 1980), pp. 21–66. See also the discussion of the structure of capitalist democracies and the formation of

demands in this system in Joshua Cohen and Joel Rogers, *On Democracy* (New York: Penguin Books, 1983), chap. 3.

15. For two summaries of this literature, see Arnold J. Heidenheimer, Hugh Heclo, and Carolyn Teich Adams, *Comparative Public Policy*, 2d ed. (New York: St. Martin's Press, 1983); Michael Shalev, "The Social Democratic Model and Beyond: Two Generations of Comparative Research on the Welfare State," Working Paper no. 8, The Bertelsmann Quality of Working Life Program, The Work and Welfare Research Institute, The Hebrew University of Jerusalem, August 1982. For focused studies that attempt to correlate policy development with the political and economic organization and strategies of labor, see Walter Korpi and Michael Shalev, "Strikes, Power and Politics in the Western Nations, 1900-1976," in Zeitlin, ed., *Political Power and Social Theory*, vol. 1, pp. 301-334; J. Rogers Hollingsworth, "The Political-Structural Basis for Economic Performance," *Annals of the AAPSS* 459 (January 1982): 28-44; E. H. Stephens and John D. Stephens, "The Labor Movement, Political Power, and Workers' Participation in Western Europe," in Zeitlin, ed., *Political Power and Social Theory*, vol. 3 (1982), pp. 215-249; and David R. Cameron, "The Politics and Economics of the Business Cycle," in Thomas Ferguson and Joel Rogers, eds. *The Political Economy* (Armonk, N.Y.: M. E. Sharpe, 1984), pp. 237-262.

16. On organizational imperatives, see Anthony Downs, *Inside Bureaucracy* (Boston: Little, Brown, 1967). Although Steven Kelman stresses the importance of the proprotectionist orientations of health and safety professionals in explaining OSHA rulemaking, his comparative account of regulation in the United States and Sweden builds from similar organizational theory assumptions. Agency staffers' protectionist values and orientations can play such a powerful role in shaping agency policy only if the agency is free to make decisions independent of the kinds of structural and political considerations noted in this book. See Steven Kelman, *Regulating America, Regulating Sweden* (Cambridge: MIT Press, 1981). As I make clear, I do not find Kelman's claim for OSHA's organizational autonomy from external pressures convincing—nor his claim that the United States and Sweden made similar decisions about occupational safety and health.

[1] The Political Economy of Workplace Regulation

1. The carcinogen cost estimate is in W. Kip Viscusi, *Risk by Choice* (Cambridge: Harvard University Press, 1983), p. 117. Viscusi converted the CWPS estimate into 1980 prices. For the noise cost estimate, see John F. Morrall III, "Exposure to Occupational Noise," in Miller and Yandle, eds., *Benefit-Cost Analyses of Social Regulation*, p. 42. Note that Morrall takes exception to

OSHA's estimate and suggests that the "gross total costs" could be as high as $27.7 billion. Morrall's estimate is based on his recalculation of information in a 1976 study done for OSHA's economic impact statement. On chemical labeling, see U.S. Department of Labor, Occupational Safety and Health Administration, *Draft Regulatory Analysis and Environmental Impact Statement for the Hazards Identification Standard* (Washington, D.C., January 1981), p. I-52. This is the present discounted value in 1980 dollars.

2. Carl Gersuny, *Work Hazards and Industrial Conflict* (Hanover, N.H.: University Press of New England, 1981), p. 2.

3. Bureau of Labor Statistics, *Occupational Injuries and Illnesses in the United States, 1980* (Washington, D.C.: GPO, 1982), table 1, pp. 2-13.

4. For an excellent discussion of this issue, see Edward S. Herman, *Corporate Control, Corporate Power* (Cambridge, England: Cambridge University Press, 1981), chap. 1.

5. See Harry Braverman, *Labor and Monopoly Capital* (New York: Monthly Review Press, 1974), chaps. 5 and 10; Richard Edwards, *Contested Terrain* (New York: Basic Books, 1979), chap. 8; Michael Burawoy, *Manufacturing Consent* (Chicago: University of Chicago Press, 1979); Dan Clawson, *Bureaucracy and the Labor Process* (New York: Monthly Review Press, 1980), chap. 2.

6. See J. K. M. Gevers, "Worker Participation in Health and Safety in the EEC: The Role of Representative Institutions," *International Labour Review* 122, no. 4 (July–August 1983): 411–428.

7. For a general discussion of health and safety in labor markets, see William T. Dickens, "Occupational Safety and Health Regulation and Economic Theory," in William Darity Jr., ed., *Labor Economics* (Boston: Kluwer-Nijhoff Publishing, 1984), pp. 133–173. On risk premiums, see Martin J. Bailey, *Reducing Risks to Life* (Washington, D.C.: American Enterprise Institute, 1980), p. 36; Robert S. Smith, *The Occupational Safety and Health Act* (Washington, D.C.: American Enterprise Institute, 1976), p. 30; Paul J. Leigh, "Compensating Wages for Employment in Strike-Prone or Hazardous Industries," *Social Science Quarterly* 65 (March 1984): 89–99; Viscusi, *Risk by Choice*, chap. 6.

8. Baruch Fischhoff, Sarah Lichtenstein, Paul Slovic, Stephen L. Derby, and Ralph L. Keeney, *Acceptable Risk* (Cambridge, England: Cambridge University Press, 1981), pp. 28–30.

9. For evidence on the existence of risk premiums, see Viscusi, *Risk by Choice*, pp. 98–106. Reservations about studies such as this one are summarized in Office of Technology Assessment, *Preventing Illness and Injury in the Workplace* (Washington, D.C.: GPO, 1985), pp. 313–314. For evidence that some of the riskiest occupations are also the lowest paid, see Norman Root and Deborah Sebastian, "BLS Develops Measure of Job Risk by Occupation," *Monthly Labor Review* 104, no. 10 (1981): 28.

10. Nicholas A. Ashford, *Crisis in the Workplace* (Cambridge: MIT Press, 1976), pp. 502–505.

11. Peter S. Barth with H. Allan Hunt, *Workers' Compensation and Work-Related Illnesses and Diseases* (Cambridge: MIT Press, 1982); and *The Report of the National Commission on State Workmen's Compensation Laws* (Washington, D.C.: GPO, 1972).

12. Asger T. Braendgaard, "Occupational Health and Safety Legislation and Working Class Action" (Ph.D. dissertation, University of North Carolina at Chapel Hill, 1974), chap. 6.

13. For a summary of worker participation systems, see Gevers, "Worker Participation," pp. 411–428. For an overview of the Western European programs in general, see L. Parmeggiani, "State of the Art: Recent Legislation on Workers' Health and Safety," *International Labour Review* 121, no. 3 (May–June 1982): 271–285; J. W. Leopold and P. B. Beaumont, "Joint Health and Safety Committees in the United Kingdom," *Economic and Industrial Democracy* 3 (1982): 263–284; A. Ian Glendon and Richard T. Booth, "Worker Participation in Occupational Health and Safety in Britain," *International Labour Review* 121, no. 4 (July–August 1982): 399–416.

14. On West Germany, see Friedrich O. Hauss and Rolf D. Rosenbrock, "Occupational Health and Safety in the Federal Republic of Germany," *International Journal of Health Services* 14, no. 2 (1984): 279–287; and Gevers, "Worker Participation," p. 421.

15. Steven Deutsch, "Work Environment Reform and Industrial Democracy," *Sociology of Work and Occupations* 8, no. 2 (May 1981): 180–194; Norman Eiger, "Economic Democracy and the Democratizing of Research: The Approach of the Swedish Labor Movement," *Labor Studies Journal* 7, no. 2 (Fall 1982): 123–141; Kelman, *Regulating America, Regulating Sweden*; and Bertil Gardell, "Scandinavian Research on Stress in Working Life," *International Journal of Health Services* 12, no. 1 (1982): 31–41.

16. On West Germany, see Hauss and Rosenbrock, "Occupational Health and Safety in Germany," pp. 279–284. For a discussion of how the public health orientation might be applied in the United States, see Frank Goldsmith and Lorin E. Kerr, *Occupational Safety and Health* (New York: Human Sciences Press, 1982), chap. 1.

[2] Before OSHA

1. Although comprehensive data on accident rates are not available prior to 1920, what evidence we do have indicates that the mechanization of work in the late nineteenth and early twentieth centuries led to significantly in-

creased accident and death rates. See David M. Gordon, Richard Edwards, and Michael Reich, *Segmented Work, Divided Workers* (Cambridge, England: Cambridge University Press, 1982), p. 148.

2. Ibid., pp. 158, 179.

3. See these two classics: Gabriel Kolko, *The Triumph of Conservatism* (Chicago: Quadrangle Books, 1967); and James Weinstein, *The Corporate Ideal in the Liberal State, 1900-1918* (Boston: Beacon Press, 1968).

4. Daniel Berman, *Death on the Job* (New York: Monthly Review Press, 1978), pp. 4-31.

5. See Crystal Eastman, ed., *Work Accidents and the Law* (New York: Charities Publication Committee, 1910), pp. 244-245; Katherine Stone, "The Origins of Job Structures in the Steel Industry," *Review of Radical Political Economics* 6, no. 2 (1974): 113-174; David S. Beyer, "Safety Provisions in the United States Steel Industry," in Eastman, ed., *Work Accidents*, pp. 344-345; and David Brody, *Steelworkers in America, The Nonunion Era* (New York: Harper & Row, 1969), pp. 166-168. U.S. Steel also took an active hand in organizing the private safety movement. Prompted by Gary, the Association of Iron and Steel Electrical Engineers established the National Safety Council in 1911; the NSC's agenda was modeled on the steel industry's experience. Berman, *Death on the Job*, pp. 21-22.

6. Weinstein, *Corporate Ideal*, chap. 2.

7. Ibid.

8. David Noble, *American by Design* (New York: Oxford University Press, 1977), pp. 80-82.

9. There were around 35,000 professionals employed in the field in the late 1960s, including toxicologists, industrial hygienists, safety engineers, doctors, and nurses. Approximately 85% of these professionals worked in the private sector, employed by individual firms, industrywide associations, private standard-setting and research organizations, and the insurance industry.

10. Data collected on a voluntary basis indicated that NSC members had injury rates that were, on the average, one-half those of their non-NSC counterparts. On the NSC, see Berman, *Death on the Job*, pp. 76-78.

11. The committee that revised the Z16.1 standard governing the reporting of accidents allotted only a single position to the AFL-CIO, 14 positions to government agencies, and 33 positions to business, insurance, and trade associations. Ibid., pp. 79-80.

12. Joseph A. Page and Mary-Win O'Brien, *Bitter Wages* (New York: Grossman, 1973), pp. 23, 82, 158-161, 199; Berman, *Death on the Job*, pp. 79, 80, 99; Department of Labor Report, in U.S. Congress, Senate, *Hearings Before the Subcommittee on Labor of the Committee on Labor and Public Welfare*, 91st Cong., 1970, p. 1445; hereinafter cited as *Senate Hearings*, 1970; ANSI

Managing Directors Report, in *Senate Hearings*, 1970, p. 440; and Ashford, *Crisis in the Workplace*, pp. 119, 154.

13. *Standards and Regulations, Report*. Department of Labor, Record Group 174, James D. Hodgson, 1970, File LL-2-3, OSH, National Archives.

14. For a fascinating view of business attitudes, see *Public Hearings on Proposed Revisions of Safety and Health Standards for Federal Supply Contracts under the Walsh-Healey Public Contracts Act*, March 1964, Department of Labor Library. Particularly instructive is the testimony of William B. Barton, U.S. Chamber of Commerce; Wesley M. Graff, National Association of Manufacturers; Wallace Smith, American Mutual Insurance Alliance; and Carl H. Hageman, Manufacturing Chemists Association. See also Berman, *Death on the Job*, pp. 76-77.

15. Goldsmith and Kerr, *Occupational Safety and Health*, p. 35.

16. Ibid., p. 31.

17. Statement of Charles Lema, safety chairman, USWA, Jersey City, N.J., March 7, 1970. *Senate Hearings*, 1970, pt. 1, p. 806. See also ORC Management, *New Dimensions for Management*, pp. 1-2, 27. The author of this memo is Leo Teplow, former official of American Iron and Steel Institute.

18. On accords in general and the postwar accord in particular, see Samuel Bowles and Herbert Gintis, "The Crisis of Liberal Democratic Capitalism: The Case of the United States," *Politics and Society* 11, no. 1 (1982): 51-93; Claus Offe, "Competitive Party Democracy and the Keynesian Welfare State: Factors of Stability and Disorganization," in Offe, *Contradictions of the Welfare State*, pp. 179-206; and Michele I. Naples, "The Structure of Industrial Relations, U.S. Mining and Manufacturing, 1930s-1980s" (paper presented at the Rutgers University Political Economy Colloquium, February 1983).

19. See Rogers, "Divide and Conquer," chap. 5.

20. See Ashford, *Crisis in the Workplace*, pp. 492-495.

21. Frank W. Lewis, "Employers Liability," *Atlantic Monthly* 103 (January 1909): 60, quoted in Weinstein, *Corporate Ideal*, p. 41.

22. Untitled paper, Department of Labor, Record Group 51, Series 61.1a, Box 232, File T2-9, p. 3. Records of the Bureau of the Budget, National Archives.

23. Ibid. The National Commission on Workmen's Compensation estimated the median percentage of wage loss replaced in temporary total disability cases to be between 40% and 44%.

24. *Federal Workmen's Compensation Bill. Statement in Explanation*, January 14, 1969, Department of Labor Record Group 174, George P. Shultz, 1969, File LL-2-3, OSH, National Archives.

25. Barth and Hunt, *Workers' Compensation*, p. 147.

26. A. Hosey and L. Ede, *A Review of State Occupational Health Legislation*, pp. 110-137, esp. p. 126, *Senate Hearings*, 1970.

27. *State Budget Allocations and Staffing for Occupational Safety, Fiscal 1968*, LSB—Division of Programming and Research, June 1969, Record Group 174, James D. Hodgson, 1970, File LL-2-3, OSH, National Archives.

28. Ibid.

29. *Occupational Safety and Health Inspection Staff*, LSB—Division of Programming and Research, Record Group 174, James D. Hodgson, 1970, File LL-2-3, OSH, National Archives.

30. Hosey and Ede, *A Review*, pp. 110-137.

31. *Status of Occupational Health Programs in State and Local Governments, January 1969*, pp. 103-109, *Senate Hearings, 1970.*

32. *State Budget Allocations, Fiscal 1968.*

33. Page and O'Brien, *Bitter Wages*, p. 73.

34. In 1966 the department reported that only 23 states "might" be interested in a grants-in-aid program if one was enacted. *Special Task Force Report on Occupational Health and Safety*, parts III-3-4, Record Group 174, Department of Labor, W. W. Wirtz, National Archives.

35. The Service Contracts Act also unequivocally granted the secretary of labor the authority to make and enforce rules and regulations apart from the states. The department estimated a workforce of 75 million workers and Department of Labor authority over almost 43 million workers. This estimate is based on Walsh-Healey Act coverage of approximately 25 million employees at one time or another each year. Other programs accounted for another 15 to 20 million workers. Note that this includes coverage of any part of the worker's day. *Current Department of Labor Responsibilities and Activities, Attachment 6*, LSB memorandum, Record Group 174, W. W. Wirtz, 1967, File LN-5-1, Staff, National Archives.

36. 54% of the staff of LSB and 48% of its budget were devoted to maritime activities. *Existing Staff Resources* and *Current Department of Labor Responsibilities and Activities*, LSB memoranda, both in Record Group 174, W. W. Wirtz, 1967, File LN-5-1, Staff, National Archives.

37. Page and O'Brien, *Bitter Wages*, p. 100.

38. *The Kirschner Report: An Evaluation of Federal-State Relationships in the Administration of Labor Standards Law*, Record Group 174, George P. Shultz, 1969, IAGLO, National Archives.

39. H. Heimann and V. Trasko, *Evolution of Occupational Health Programs in State and Local Governments*. U.S. Congress, Senate, *Hearings on the Occupational Safety and Health Act of 1968 (S. 2864) Before the Subcommittee on Labor of the Committee on Labor and Public Welfare*, 90th Cong., 2d sess., 1968, pp. 474-475. Hereinafter cited as *Senate Hearings, 1968. Federal-State Agreements to Administer Walsh-Healey Act Safety Provisions*, Record Group 174, George P. Shultz, 1969, Committee, IAGLO, National Archives.

40. Page and O'Brien, *Bitter Wages*, p. 100.

41. Statement of John D. Hanlan. U.S. Congress, House, *Hearings on Occupational Safety and Health (on H.R 14816) Before the Select Committee on Labor of the Committee on Education and Labor*, 90th Cong., 2d sess., 1968, p. 472. Hereinafter cited as *House Hearings*, 1968.

42. Interview with Dr. Norton Nelson, July 22, 1983.

43. *Injury Rates, New York State Industries, 1962*. New York State Department of Labor, Division of Research and Statistics, Publication B-145. Reprinted in *House Hearings*, 1968, p. 95.

44. U.S. Bureau of the Census, *Historical Statistics of the United States, Colonial Times to 1970*, Bicentennial ed., pt. 1 (Washington, D.C.: GPO, 1975).

45. For the first statement of this relationship, see Max D. Kossoris, "Industrial Injuries and the Business Cycle," *Monthly Labor Review* 46 (1938): 579–594; for a contemporary account, see Robert S. Smith, "Intertemporal Changes in Work Injury Rates," Proceedings of the Industrial Relations Research Association, 1973, pp. 167–174; for a summary of this literature, see Office of Technology Assessment, *Preventing Illness*, chap. 2.

46. Smith, *Occupational Safety and Health Act*, pp. 5–9.

47. Michel Aglietta, *A Theory of Capitalist Regulation* (London: New Left Books, 1979), pp. 116–122.

48. Alexander Cockburn and James Ridgeway, "The U.S. Crisis," *Village Voice*, October 5, 1982, p. 23.

49. *Senate Hearings*, 1970, p. 1525.

50. Harvey J. Hilaski and Chao Ling Wang, "How Valid Are Estimates of Occupational Illness?" *Monthly Labor Review* 105, no. 8 (1982): 27–35.

51. Wallick's efforts resulted in *The American Worker: An Endangered Species* (New York: Ballantine, 1972). Brodeur's articles were later published as *Expendable Americans* (New York: Viking, 1973). Selikoff's research is recounted in Statement of Dr. Irving J. Selikoff, in *House Hearings*, 1968, pp. 349–355. Davidson's research appeared in *Peril on the Job* (Washington, D.C.: Public Affairs Press, 1970).

52. "New Data on Asbestos Indicate Cover-up of Effects on Workers," *Washington Post*, November 12, 1978, p. A1.

53. Office of Technology Assessment, *Preventing Illness*, p. 37.

[3] The Origins of the OSH Act

1. Jack Barbash, "The Causes of Rank and File Unrest," in Joel Seidman, ed., *Trade Union Government and Collective Bargaining* (New York: Praeger, 1970), p. 52; P. K. Edwards, *Strikes in the United States, 1881–1974* (New York: St. Martin's Press, 1981), p. 183.

2. Robert P. Quinn and Linda J. Shepard, *The 1972–73 Quality of Employment Survey, Descriptive Statistics with Comparison Data from the 1969–1970 Survey of Working Conditions* (Ann Arbor: Survey Research Center, Institute for Social Research, University of Michigan, 1974), pp. 149–153.

3. Ralph Nader, "The Violence of Omission," *Nation* 208 (February 10, 1969): 166–169; "Coal Workers Strike West Virginia Mines in Health-Aid Fight," *Wall Street Journal*, February 21, 1969, p. 3; "West Virginia Senate Fails to Satisfy Goals Set by Striking Miners," *Wall Street Journal*, March 6, 1969, p. 21; "West Virginia Miners Accelerate Campaign over 'Black Lung' Bill," *Wall Street Journal*, March 7, 1969, p. 2; "West Virginia Miners Expected Back on Jobs Within Next Few Days," *Wall Street Journal*, March 10, 1969, p. 12; "Labor Letter," *Wall Street Journal*, March 11, 1969, p. 1; "Success in 'Black Lung' Aid Fight Spurs Coal Miner Drive for More Health Gains," *Wall Street Journal*, April 1, 1969, p. 4; "New Strike Threatened in Coal Mines of West Virginia over 'Black Lung' Law," *Wall Street Journal*, June 30, 1969, p. 4; "Cotton Mills Workers Said to Risk Lung Ill of a 'Different Color,'" *Wall Street Journal*, October 30, 1969, p. 17; "Labor Letter," *Wall Street Journal*, April 7, 1970, p. 1.

4. Personal interview with Jack Sheehan, March 4, 1983.

5. U.S. Congress, House, *Hearings on H.R. 843, H.R. 3809, H.R. 4294, and H.R. 13373 Before the Select Subcommittee on Labor of the Committee on Education and Labor*, 91st Cong., 1st sess., 1969, p. 652. Hereinafter cited as *House Hearings, 1969*.

6. Interview with Dr. Nelson, July 22, 1983; interview with George Taylor, February 10, 1983. Cover letter to Dr. William H. Stewart, Surgeon General, from Dr. Norton Nelson, November 19, 1965, transmitting *Frye Report*. PHS/DHEW, 1965.

7. Public Health Service, *Protecting the Health of 80 Million Americans* (Washington, D.C.: GPO, 1965), pp. 9, 13, 17.

8. See letter to Secretary W. Willard Wirtz from F. C. Love, executive vice-president, Kerr-McGee Corporation, June 7, 1967, Record Group 174, Department of Labor, W. Willard Wirtz, 1967, File LN-5, Occupational Health and Safety, National Archives.

9. *Summary of Events Relating to Uranium Miner Problems, January 27, 1969*, Record Group 174, George P. Shultz, 1969, Legislation, File LL-2-3, Occupational Health and Safety, National Archives; *Memorandum for the Secretary. From Esther Peterson. Subject: Background on Uranium Mining Hazards*, n.d., Record Group 174, Department of Labor, W. Willard Wirtz, 1967, File LN-5, OSH, National Archives.

10. Ibid.

11. Memo from Esther Peterson to W. W. Wirtz, 1967, Record Group 174, W. W. Wirtz, National Archives.

12. Interview with Anthony Mazzochi, March 9, 1983; interview with Stanley Aronowitz, July 25, 1983.

13. Barbash, "Causes of Rank and File Unrest"; Edwards, *Strikes*, pp. 195ff.; Philip Taft, "Rank and File Unrest in Historical Perspective," in Seidman, ed., *Trade Union Government*, pp. 97–101.

14. Donald G. Sofchalk, "United Steelworkers of America," in Gary M. Fink, ed., *Labor Unions* (Westport, Conn.: Greenwood Press, 1977), pp. 357–359. "Steelworkers Reach Agreement on Proposal to Protect Earnings," *Wall Street Journal*, August 8, 1969, p. 4; "Bid to Oversee Balloting of Steelworkers Denied," *Wall Street Journal*, February 5, 1969, p. 20; "Labor Letter," *Wall Street Journal*, June 17, 1969, p. 1.

15. "Steel Union Girds for Record Settlement in 1971 Bargaining, Abel Tells Convention," *Wall Street Journal*, September 29, 1970, p. 3; "Labor Letter," *Wall Street Journal*, November 24, 1970, p. 1.

16. Letter from George Meany to W. Willard Wirtz, December 5, 1963; letter to George Meany from W. Willard Wirtz, December 18, 1963; letter to W. Willard Wirtz from Andrew J. Biemiller, December 19, 1963; letter to Andrew J. Biemiller from W. Willard Wirtz, February 3, 1964; letter to W. Willard Wirtz from Andrew J. Biemiller, April 3, 1964; letter to Andrew J. Biemiller from W. Willard Wirtz, April 17, 1964. All Record Group 174, Department of Labor, W. Willard Wirtz, 1963, AFL–CIO; 1964, AFL–CIO; National Archives.

17. Staffers on the Senate Labor Committee despaired of attracting television reporters to their hearings because worker health and safety was not "sexy." In response, they carefully orchestrated Ralph Nader's appearances in support of the Democratic bill. Even then, their effort failed to capture and hold the media's interest. Interview with Robert Nagle, July 29, 1983. For an analysis of the media and OSHA, see Patrick Gerald Donnelly, "Social Problems Versus Social Issues: The Case of Worker Safety and Health" (Ph.D. dissertation, University of Delaware, June 1981), pp. 95–106.

18. Mazzochi credits environmentalists with showing him how worker health raised larger environmental issues, particularly issues about the control of industrial technology. Interview with Tony Mazzochi, March 9, 1983.

19. "Labor Notes," *Wall Street Journal*, April 23, 1968, p. 1; letter from Peter Borelli to all senators, November 4, 1970; letter from Environmental Action to all senators, October 7, 1970. The letter was signed by a dozen scientists and environmentalists, including George Wald of Harvard, Dr. Paul Cornely, president of the American Public Health Association, Dr. Samuel S. Epstein of the Children's Cancer Research Foundation, Gary Soucie of the Friends of the Earth, Rene Dubos of Rockefeller University, Paul Ehrlich of Stanford, and Dr. Mary Bunting of Radcliffe.

20. "Nader Recruits 80 New Student 'Raiders' to Investigate the Operations

of a Dozen Federal Agencies," *New York Times*, June 1, 1969, p. 43. Letter to Secretary Shultz from Ralph Nader, May 22, 1969, Record Group 174, George P. Shultz, 1969, Labor Standards, File LN-5, OSH, National Archives. "Labor Letter," *Wall Street Journal*, December 30, 1969, p. 1. The Page and O'Brien study, *Bitter Wages*, is the fruit of the Nader effort.

21. Walter A. Rosenbaum, *The Politics of Environmental Concern*, 2d ed. (New York: Praeger, 1977), pp. 76–81; Philip P. Micklin, "Water Quality: A Question of Standards," in Richard A. Cooley and Geoffrey Wandesforde-Smith, eds., *Congress and the Environment* (Seattle: University of Washington Press, 1970), pp. 130–147; M. Kent Jennings, "Legislative Politics and Water Pollution Control, 1956–61," in Frederick N. Cleaveland and associates, *Congress and Urban Problems* (Washington, D.C.: Brookings Institution, 1969), pp. 72–109.

22. Everett Carll Ladd, Jr., with Charles D. Hadley, *Transformations of the American Party System*, 2d ed. (New York: Norton, 1978), chaps. 4 and 5.

23. Secretary of Labor Wirtz reported that the Johnson White House considered this to be one of the great virtues of the OSH Act. Interview with Secretary Wirtz, July 25, 1983.

24. See "Labor Leaders Alarmed by Diminishing Support Among Rank and File," *Wall Street Journal*, July 6, 1967, p. 1.

25. The source of this account appears to be Page and O'Brien, *Bitter Wages*, pp. 137–138. Meier also relies on this account; see Kenneth J. Meier, *Regulation* (New York: St. Martin's Press, 1985), p. 206. See also Steven Kelman, "Occupational Safety and Health Administration," in James Q. Wilson, ed., *The Politics of Regulation* (New York: Basic Books, 1980), p. 239.

26. Jones's account of "speculative augmentation" gives the impression that Congress got carried away on issues of this sort and stepped into areas in which it had little knowledge or expertise. See Charles O. Jones, *Clean Air* (Pittsburgh: University of Pittsburgh Press, 1978), chap. 7.

27. Califano organized a series of interdepartmental task forces to generate ideas. Reversing its traditional role as a screening agency, the Bureau of the Budget coordinated the search for legislation. Interview with Joseph Califano, February 6, 1984; interview with Charles Schultze, February 6, 1984.

28. Office of the Federal Register, *Public Papers of the Presidents of the United States, Lyndon B. Johnson, 1966* (Washington, D.C.: GPO, 1967), p. 540.

29. Testimony of W. Willard Wirtz, secretary of labor. *House Hearings*, 1968, pp. 12–18.

30. "Life or Death for Your Business?" *Nation's Business* 56, no. 4 (April 1968).

31. Henry Ford II, "The Revolution in Public Expectations," *Public Relations Journal* 26 (October 1970): 16–18.

32. Sol Linowitz, "The Growing Responsibility of Business in Public Affairs," *Management Review*, September 1966, pp. 52–55.

33. Statement of Leo Teplow, vice-president, American Iron and Steel Institute, June 19, 1968. *Senate Hearings*, 1968, pp. 348–350.

34. Statement of John O. Logan, Universal Products, June 12, 1968. *Senate Hearings*, 1968, pp. 254–257.

35. Statement of David Goldstein, M.D., June 24, 1968. *Senate Hearings*, 1968, p. 452.

36. Statement of Leo Teplow. *Senate Hearings*, 1968, pp. 348–350.

37. For the NSC position, see prepared statement of Howard Pyle, president, NSC. *Senate Hearings*, 1968, pp. 519–526.

38. Special Task Force on Occupational Safety, 1966, Record Group 174, File III-3, 8, Secretary of Labor W. W. Wirtz, National Archives. Personal communication from Gardner, August 1, 1983; interview with George Taylor, February 10, 1983; interview with Dr. Philip Lee, February 17, 1984.

39. Memorandum to John Gardner from Joseph Califano, September 23, 1966. Lyndon Baines Johnson Library, Austin, Texas. Memorandum from Health and Welfare Division (Terry Davies) to Director, Bureau of the Budget, November 3, 1966. Subject: Report on Task Force on Accident Prevention. Record Group 51, Records of the Bureau of the Budget, Series 61.1b, Box 10, Accident Prevention, National Archives.

40. The second task force was formed and chaired by the Office of Science and Technology, but officials from the PHS and LSB remained actively involved. The final report made few substantive recommendations. It was, in Califano's words, "a bomb." Memorandum from Health and Welfare Division (Terry Davies). Memo from Terry Davies to director of BOB, November 3, 1966. Subject: Report on Task Force on Accident Prevention. Record Group 51, BOB, Series 61.1b, Box 10, Accident Prevention, National Archives. Memo from Bill Carey to Director, BOB, Notes on the Califano Meeting with Task Force on Accident Prevention, Record Group 51, Records of the BOB, Series 61.1b, Box 10, Accident Prevention, National Archives. Memo for Joseph Califano from Philip R. Lee, M.D., assistant secretary for health and scientific affairs, HEW. Subject: Special Task Force Report on Occupational Health and Safety, December 14, 1966; memo for Joseph Califano from Ivan L. Bennett, Jr., deputy director, Office of Science and Technology, December 14, 1966. Both in Record Group 51, Records of the BOB, Series 69.2, Box 315, File T2-9, National Archives. Memo from Charles Elkins to Irving J. Lewis, September 11, 1967, Record Group 51, Series 68.2, Box 315, File T2-9, National Archives. Memo to Dr. Milch and Messrs. Radley, MacRaie, Messner, and Jasper from Human Resources Program Divisions, C. Elkins. Subject: Occupational Health and Safety, September 15, 1967. Record Group 51, BOB, Series 69.2, Box 315, File T2-9, National Archives. Memo for Mr. Califano from Charles L. Schultze, director, Bureau of the Budget, October 21, 1967, Record Group 51, Series 60.3a, Folder BOB/HRP/NIOSH/etc., National Archives. Schultze notes that

the secretaries had not been brought into the picture at that time. When interviewed for this study, Secretary Wirtz, Assistant Secretary Peterson, and David Swankin, director of the Bureau of Labor Standards, reported that they had been unaware of the White House's efforts during this period. Swankin believed that the Department of Labor had taken the issue of occupational safety and health to the White House in 1967, defended its claims against those of HEW, and convinced Califano to make this a priority issue. Personal interview with Secretary of Labor W. Willard Wirtz; personal interview with Assistant Secretary of Labor Esther Peterson; personal interview with David Swankin, director of the Bureau of Labor Standards, July 27–29, 1983.

41. That the meeting took place and the conversations were reported is evident in references to two memos in a variety of documents that passed between Labor and the White House. See Memo from Secretary Wirtz to Douglas Cater, White House, re: Insurance Company Representatives Meeting; memo from Secretary Wirtz to the President, September 9, 1967, re: Meeting with Insurance Company Representatives. Neither memo is in Wirtz's files or the Johnson Library at Austin. However, insurance company opposition to reform of workers' compensation was well known, and was confirmed in personal interviews with Wirtz on July 27, 1983, and with Monroe Berkowitz, who chaired a White House Task Force on Workmen's Compensation Reform, July 20, 1983; see also Untitled report on workmen's compensation attached to November 14, 1967, memo from Wirtz to Joseph Califano. All of the above are located in Record Group 174, Labor, W. Willard Wirtz, 1967, Box: White House—President, National Archives.

42. Memo for Mr. Joseph Califano, from Gardner Ackley, chairman of the Council of Economic Advisers, November 22, 1967. Subject: Labor Department Proposals on Workmen's Compensation. Folder: Task Force on Workmen's Compensation, Container 41 (1743), Lyndon Baines Johnson Library, Austin, Texas. Memo for Mr. Califano from Phillip S. Hughes, deputy director, November 28, 1967. Subject: Comments on the Department of Labor's Workmen's Compensation Paper. Record Group 51, BOB, Series 69.2, Box 315, File T2-9. National Archives. Memo to Joe Califano from Jim Gaither re: Meeting on Workmen's Compensation, November 30, 1967 (memo dated November 29, 1967). Folder: Task Force on Workmen's Compensation, Container 41 (1743), Lyndon Baines Johnson Library, Austin, Texas.

43. Suggestions for Early Action, Consideration, or Pronouncement. A Report to the President-Elect. Submitted by Arthur F. Burns, chairman, Program Coordination Group, January 6, 1969. Memorandum to Dr. Arthur Burns from Alexander P. Butterfield, February 12, 1969. Subject: Notes from the President. Both in Record Group 174, George P. Shultz, 1969, White House—Reports Requested for the President, National Archives. Memorandum for George P.

Shultz from Richard Nixon, February 11, 1969, Record Group 174, George P. Shultz, 1969, White House—Reports Requested for the President, National Archives. Interview with John Hodgson, August 10, 1983; interviews with Anthony Obadal, February 11, 1983, and July 26, 1983.

44. Ceremony for signing of Occupational Safety and Health Act of 1970, December 29, 1970, Conference Room B, Labor Department Auditorium, Record Group 174, James D. Hodgson, 1970, File LL-2-3, OSH, National Archives.

45. These meetings were described in interviews with Dr. Norton Nelson, who had chaired the working group that had supervised the preparation of the Frye Report, Dr. James Sterner, a member of the PHS's Council on Environmental Health, and George Taylor of the AFL-CIO.

46. Statement of Howard Pyle. *Senate. Hearings,* 1970, pp. 561-576.

47. Department of Commerce position on Federal Occupational Health and Safety Regulations, attached to letter from Maurice Stans to George P. Shultz, March 7, 1969, Record Group 174, George P. Shultz, 1969, Labor Standards, File LN-5, OSH, National Archives.

48. Memo. Subject: Item-by-Item Reply to the Points Raised in Secretary Stans's March 7 Letter on Occupational Safety. Attached to letter from George P. Shultz to Maurice P. Stans, March 24, 1969. Record Group 174, George P. Shultz, 1969, Labor Standards, File LN-5, OSH, National Archives.

49. This account of the informal negotiations draws heavily on private notes taken by USWA lobbyist John J. Sheehan. These notes are located in the USWA offices in Washington, D.C. and are far and away the most detailed record of the legislative history. They shed considerable light on events reported sketchily in the press at the time. These are hereinafter referred to as the Sheehan Papers. I have checked Sheehan's account against press accounts, other actors' recollections, and the documentary record. See also "Battle Looms over Safety Bill," *Washington Post,* April 20, 1970, p. A4; "Omnibus Safety Bill—Congressional Battle Reaches a Climax," *Occupational Hazards,* July 1970, pp. 41-44.

50. "Dispute Mires Job Safety Bill," *Washington Post,* October 1, 1970, p. G1.

51. See memo from Leo Teplow, "Desirable Objectives in Federal Safety Legislation," May 18, 1970. Sheehan Papers.

52. The act requires the heads of federal agencies to set up and operate programs that are consistent with the standards established by the secretary of labor (Sec. 19a). The act also provides that state and local government employees be covered under plans submitted by each state to the "extent permitted by its law" (Sec. 18c[6]).

53. Union lobbyists were willing to include the term "technical feasibility," but industry lawyers advised the industry lobbyists that they stood a better

chance of getting the courts to interpret "feasibility" broadly to include costs if the word was unqualified. As a result, the act was written without a more specific guideline as to how feasibility was to be taken into account. Interview with John J. Sheehan, June 8, 1983; interview with Leo Teplow, February 11, 1983; interviews with Anthony Obadal, February 11, 1983, and July 26, 1983.

54. U.S. Congress, Senate Committee on Labor and Public Welfare, Subcommittee on Labor, *Legislative History of the Occupational Safety and Health Act of 1970* (Washington, D.C.: GPO, 1971), p. 986.

55. Ibid., p. 367.

56. Interview with Frank Barnako, July 25, 1983.

57. Testimony of Ralph Nader. *Senate Hearings*, 1968, pp. 610–611.

[4] The Politics of Deregulation

1. In general, see David Vogel, "The Power of Business in the United States: A Re-appraisal," *British Journal of Political Science* 13 (1983): 19–43; Michael Useem, *The Inner Circle* (New York: Oxford University Press, 1984), chap. 6; and Sar A. Levitan and Martha R. Cooper, *Business Lobbies* (Baltimore: Johns Hopkins University Press, 1984).

2. David Rockefeller, "Corporate Task for 70's: Social Action, Not Words," *Commercial and Financial Chronicle* 215, no. 7166 (1972): 69.

3. Levitan and Cooper, *Business Lobbies*, pp. 19, 24, 27–33; David Noble and David Dickson, "By Force of Reason," in Thomas Ferguson and Joel Rogers, eds., *The Hidden Election* (New York: Pantheon, 1981), p. 270; and "The Naderites of the Other Side," *New York Times*, September 30, 1979, p. F7; "OSHA Policy Could Cost $88 Billion, Raise Inflation Rate," AIHC Estimates, *Occupational Safety and Health Reporter*. Hereinafter cited as *OSH Reporter*, March 30, 1978, pp. 1636–1637.

4. "Naderites of the Other Side."

5. Michael S. Brown, "Setting Occupational Health Standards: The Vinyl Chloride Case," in Dorothy Nelkin, ed., *Controversy: Politics of Technical Decisions*, 2d ed. (Beverly Hills, Calif.: Sage, 1984), pp. 125–142, esp. pp. 128–132.

6. The industry actually prospered after the OSHA standard forced it to adopt new, more efficient production methods. On the original industry estimate, actual costs and the SPI response, see "Did Industry Cry Wolf?" *New York Times*, December 28, 1975, p. F1; "Economic Impact of Proposed Vinyl Chloride Rules Studied by A. D. Little," *OSH Reporter*, June 27, 1974, pp. 86–87; "Obey Hits Warnings on Cost of VC Rule; SPI Says Wait and See," *OSH Reporter*, July 31, 1975, pp. 295–296.

7. *Industrial Union Department* v. *Hodgson*, 499 F.2d 467 (D.C. Cir. 1974).

8. Ibid.

9. *American Textile Mfrs. Inst.* v. *Bingham*, No. 78-1378 (4th Cir. 1978), 212; "Industry Claims Lead Standard 'Classic Example of Agency Capriciousness,'" *OSH Reporter*, June 21, 1979, p. 62; "OSHA Policy Could Cost $88 Billion," pp. 1636–1637.

10. Mark A. de Bernardo, *OSHA, Are You Listening?* Staff Report, Labor Relations Committee, Chamber of Commerce of the U.S. (Washington, D.C.: 1979).

11. Arthur Andersen & Co., "Executive Summary," in *Cost of Government Regulation Study for the Business Roundtable*, 3 vols. (New York: March 1979).

12. Ibid.

13. "Hazard Assessment Must be Made Outside Political Arena, Lang Says," *OSH Reporter*, July 5, 1979, p. 105.

14. "OSHA Policy Could Cost $88 Billion," pp. 1636–1637.

15. NAM Occupational Safety and Health Committee, *Recommendations for OSHA Reform*, submitted to the agency, April 24, 1981; letter to Thorne G. Auchter from C. Neil Norgen, chairman, Occupational Safety and Health Committee, NAM, April 28, 1981; statement of Richard Lesher, president, NAM. All from U.S. Congress, Senate, *Oversight Hearings Before Senate Labor and Human Resources Committee*, March–April 1980.

16. "Objection Raised to Task Force Proposals on Tax Incentives, Penalty Modifications," *OSH Reporter*, March 29, 1979, pp. 1606–1607. Statement by the Chamber of Commerce to Senate Select Committee on Small Business/Governmental Affairs, July 30, 1980; statement of Richard Lesher; NAM, *Recommendations for OSHA Reform*; and letter to Thorne Auchter from C. Neil Norgen.

17. Statement on OSHA Oversight Before the Senate Labor and Human Resources Committee for the Chamber of Commerce of the United States, April 1, 1980; typescript. NAM, *Recommendations for OSHA Reform*; letter to Thorne G. Auchter from C. Neil Norgen; statement of Richard Lesher.

18. William R. Greer, "Value of One Life?" *New York Times*, June 26, 1985, p. A1.

19. Russell Franklin Settle, "The Welfare Economics of Occupational Safety and Health Standards" (Ph.D. dissertation, University of Wisconsin, 1974).

20. Jacqueline Karnell Corn and Morton Corn, "The Myth and the Reality," in Robert F. Lanzillotti, ed., *Economic Effects of Government-Mandated Costs* (Gainesville: University Presses of Florida, 1977), p. 106.

21. "Regulatory Reform Legislation," recommendations of the Business Roundtable, January 8, 1980, mimeo.

22. Murray Weidenbaum and Robert DeFina, *The Cost of Federal Regulation of Economic Activity* (Washington, D.C.: American Enterprise Institute, 1978). The 1979 figures are taken from John E. Schwarz, *America's Hidden Success* (New York: Norton, 1983), p. 99.

23. Testimony of Murray L. Weidenbaum, "Use of Cost-Benefit Analysis by Regulatory Agencies." U.S. Congress, House, *Joint Hearings Before the Subcommittee on Oversight and Investigations and the Subcommittee on Consumer Protection and Finance of the Committee on Interstate and Foreign Commerce.* 96th Cong., 1st sess., 1979, pp. 319–323.

24. See William K. Tabb, "The Cost of Not Regulating: The Political Economy of Protecting Workers, Consumers, and Our Living Space" (manuscript, n.d.); Green and Waitzman, *Business War on the Law;* Schwarz, *America's Hidden Success,* pp. 101–106.

25. Testimony of Murray L. Weidenbaum, p. 452.

26. Ibid., pp. 321–324.

27. Zeckhauser and Nichols, *OSHA,* pp. 163–167, 188–191.

28. Smith, *OSH Act,* p. 2.

29. James Robert Chelius, *Workplace Safety and Health* (Washington D.C.: American Enterprise Institute, 1977), pp. 2–4.

30. Zeckhauser and Nichols, *OSHA,* p. 165.

31. Ibid., p. 167.

32. Viscusi, *Risk by Choice,* pp. 80–81.

33. Aaron Wildavsky, "Richer Is Safer," *Public Interest* 60 (Summer 1980): 23, 29.

34. Ibid., p. 33.

35. Ibid.

[5] Labor's Defense of Social Regulation

1. Bertil Gardell and Bjorn Gustavsen, "Work Environment Research and Social Change: Current Developments in Scandinavia," *Journal of Occupational Behavior* 1 (January 1980).

2. On long-standing opposition to greater regulation, but support for social regulation, see Seymour Martin Lipset and William Schneider, "The Public View of Regulation," *Public Opinion* 2, no. 1 (January/February 1979): 6–13. On the differences between support for economic regulation and protective regulation, see Louis Harris and Associates, *Consumerism in the Eighties,* a poll conducted for Atlantic Richfield Company, 1983, pp. 35–36, 38.

3. Cambridge Reports, *Public and Worker Attitudes Toward Carcinogens and Cancer Risk* (Cambridge, Mass.: Cambridge Reports, 1978).

4. Cambridge Reports, *Cambridge Report*, no. 16 (Cambridge, Mass.: Cambridge Reports, 1978).

5. "Toxic Chemical Dumps: Corrective Action Desired," *ABC News–Harris Survey* 11, no. 82 (July 7, 1980): 3.

6. Louis Harris, "Substantial Majorities Indicate Support for Clean Air and Clean Water Acts," *The Harris Survey* 47 (June 11, 1981): 2–3.

7. Cambridge Reports, *Public and Worker Attitudes*.

8. "More Conservatives Share 'Liberal' View," *New York Times*, January 22, 1978, pp. A1ff. The poll is on p. 30.

9. Louis Harris, "Views on Government Regulation of Business," *The Harris Survey* 64 (August 10, 1981): 3.

10. Harris and Associates, *Consumerism in the Eighties*, p. 46; quote is on p. 38.

11. Philip Shabecoff, "Poll Finds Majority Opposes Weaker Environment Laws," *New York Times*, November 14, 1982, p. A74.

12. Harris and Associates, *Consumerism in the Eighties*, pp. 46–47, 65, 70–71.

13. Morris E. Davis, "The Impact of Workplace Health and Safety on Black Workers: Assessment and Prognosis," *Labor Studies Journal* 6, no. 1 (Spring 1981): 29–40.

14. Jeanne H. Stellman, "Occupational Health and Women Workers: A Review," *Labor Studies Journal* 6, no.1 (Spring 1981): 16–28. Norman Root and Judy R. Daley, "Are Women Safer Workers? A New Look at the Data," *Monthly Labor Review* 103, no. 9 (1980): 3–10.

15. Richard B. Freeman and James L. Medoff, *What Do Unions Do?* (Cambridge: Harvard University Press, 1984), pp. 199–201.

16. The 1969 and 1972 data are comparable and indicate a small increase. The 1977 data are not comparable with data from the previous surveys, but the increase is significant. Richard Frenkel and W. Curtiss Priest, *Health, Safety and the Worker* (Cambridge, Mass.: Center for Policy Alternatives, 1979), pp. 78, 93–94.

17. Ibid., pp. 179, 181, 214.

18. Bureau of National Affairs, *OSHA and the Unions* (Washington, D.C.: The Bureau, 1973), p. 1. See also Winston Tillery, "Safety and Health Provisions Before and After OSHA," *Monthly Labor Review* 98, no. 9 (1975): 40–43.

19. BNA, *OSHA and the Unions*, pp. 3–4.

20. Health Research Group, *Survey of Occupational Health Efforts of Fifteen Major Labor Unions* (Washington, D.C.: Health Research Group/Public Citizen, 1976); Michael Merrill, "Organizing for Job Safety and Health," *So-*

cialist Review 66 (November–December 1982): 133–138; Dan MacLeod, United Auto Workers, *What Every UAW Representative Should Know About Health and Safety* (Detroit: UAW, 1979).

21. Health Research Group, *Survey*; Public Citizen Health Research Group, *1983 Survey of Fourteen Union Occupational Safety and Health Programs* (Washington, D.C.: Public Citizen Health Research Group 1984).

22. "Abel to IUD Safety Confab: No 'Environmental Blackmail,'" *Steel Labor*, July 1971, p. 3.

23. Richard Grossman, "Environmentalists and the Labor Movement," *Socialist Review* 15, nos. 4 and 5 (1985): 63–87. See also Richard Kazis, *Fear at Work* (New York: Pilgrim Press, 1982), prepared under the auspices of EFFE; and Industrial Union Department, "OSHA/Environmental Watch" (Washington, D.C.: AFL–CIO), published four times yearly.

24. "Lax Enforcement Perils Health and Safety Legislation," *Labor Today*, January 1973, p. 10. My discussion of the COSH movement is based on interviews with COSH activists and the following sources: Daniel Berman, "Grassroots Coalitions in Health and Safety: The COSH Groups," *Labor Studies Journal* 6, no. 1 (Spring 1981): 104–113; Merrill, "Organizing for Job Safety and Health"; and Gail Robinson, "Organizing Against Workplace Pollution," *Environmental Action*, September 1980, pp. 4–10.

25. OSHA gave 40 awards totaling $3.4 million in 1977; 45 awards totaling $3 million in 1978; $6.4 million in awards in 1979; and 66 awards totaling $3.5 million in 1980. *The President's Report on Occupational Safety and Health* (Washington, D.C.: U.S. Department of Labor, 1978, 1979, 1980), Appendix C.

26. *Do-It-Yourself Tactics: Local Action on Job Safety* (Philadelphia: PHILAPOSH and American Labor Education Center, n.d.).

27. See "One Man Scoops the Experts," *In These Times*, March 18–31, 1981; "Fighting for Their Lives," *Solidarity*, pp. 7–9.

28. Legislative Department, USWA, "Safety and Health Legislation, 1978–1979," p. 2.

29. MacLeod, *What Every UAW Representative Should Know*, p. 17.

30. Public Citizen Health Research Group, *Survey*, p. 1.

31. Statement of Lane Kirkland, president of the AFL–CIO. U.S. Congress, Senate, *Hearings of the Committee on Labor and Human Resources on Oversight of the Occupational Safety and Health Act*, April 1, 1980. Testimony, AFL–CIO. U.S. Congress, Senate, *Hearings of the Subcommittee on Oversight and Investigation, Subcommittee on Labor, Labor and Human Resources Committee*, September 23, 1981.

32. Grossman, "Environmentalists and the Labor Movement," p. 66.

33. Ibid., p. 76.

34. Testimony, AFL–CIO.

35. Berman, "Grassroots Coalitions in Health and Safety"; Merrill, "Organizing for Job Safety and Health"; Robinson, "Organizing Against Workplace Pollution."

36. Rob Wrenn, "The Decline of American Labor," *Socialist Review* 15, nos. 4 and 5 (1985): 89–117.

37. Wrenn, "The Decline of American Labor," pp. 97–103.

38. Freeman and Medoff, *What Do Unions Do?*, pp. 194–198.

[6] The White House Review Programs

1. "Inflation Boosts Chance for Policy Changes," *National Journal*, September 9, 1974, p. 1394.

2. Weidenbaum and DeFina, *Cost of Federal Regulation*, p. 3.

3. Edward F. Denison, *Accounting for Slower Economic Growth* (Washington, D.C.: Brookings Institution, 1979), pp. 69, 71, 129, 145; and Edward F. Denison, "Effects of Selected Changes in the Institutional and Human Environment upon Output Per Unit of Input," *Survey of Current Business* 58, no. 2 (1978): 21.

4. George C. Eads and Michael Fix, *Relief or Reform? Reagan's Regulatory Dilemma* (Washington, D.C.: Urban Institute, 1984), pp. 37, 46–50.

5. See the Data Resources study for the Council on Environmental Quality, "The Macroeconomic Impact of Federal Pollution Control Programs, 1978 Assessment," prepared for the Council on Environmental Quality and the Environmental Protection Agency. Submitted January 11, 1979.

6. Worker health and safety absorbed 12.3% of capital spending in the steel industry in 1972. Economics Department, *Annual McGraw-Hill Survey of Investment in Employee Safety and Health* (New York: McGraw-Hill, 1984), table iv.

7. "Marshall Supports Effort by Controversial Task Force," *OSH Reporter*, July 21, 1977, pp. 235–236.

8. "Carter's Assault on the Costs of Regulation," *National Journal*, August 12, 1978, p. 1282.

9. Congressional Quarterly, *Federal Regulatory Directory, 1981–1982* (Washington, D.C.: Congressional Quarterly, 1981), p. 73.

10. Executive Order 11, 821, November 27, 1974.

11. "Inflation Boosts Chance for Policy Changes," p. 1394; "Ford Initiates Reform Offensive," *National Journal*, July 5, 1975, p. 1000; "President Raises Cost-Benefit Issue for Safety, Noise Regulations," *OSH Reporter*, May 1, 1975, p. 1559. Ford's figure, based on a 1975 OMB study, was soon abandoned by

White House staffers and a General Accounting Office study ultimately discredited it. Eads and Fix, *Relief or Reform?*, p. 28. See also, Julius W. Allen, *Estimating the Costs of Federal Regulation* (Washington, D.C.: Congressional Research Service, 1978).

12. *Washington Post* interview with Ronald Reagan, June 5, 1980, quoted in Sierra Club, *Poisons on the Job*, Natural Heritage Report no. 4 (San Francisco: Sierra Club, 1982), p. 21.

13. Ronald Reagan, "Address to Joint Session of Congress on the Economy," in *Congressional Quarterly Weekly Report*, February 21, 1981, p. 363.

14. "Vice President Bush's Statement on the Executive Order on Regulatory Management Signed by President Reagan, February 17, 1981"; typescript.

15. Ibid.

16. "Another Round of Reform," *National Journal*, November 27, 1976, p. 1712.

17. "Carter Has Landed Running on Regulatory Reform Issues," *National Journal*, April 16, 1977, pp. 59–63; David Dickson and David Noble, "By Force of Reason: The Politics of Science and Technology Policy," in Ferguson and Rogers, eds., *Hidden Election*, p. 269.

18. "President Endorses Substitution of Economic Incentives for Safety Rules," *OSH Reporter*, July 28, 1977, p. 267.

19. The task force included the secretaries of the Departments of Commerce, Treasury, and Labor, the chair of the CEA, the director of OMB, the attorney general, and the president's domestic policy adviser.

20. Charles E. Ludlam, *Undermining Public Protections: The Reagan Administration's Regulatory Program* (n.p., Alliance for Justice, 1981), p. 9.

21. General Accounting Office, *Improved Quality, Adequate Resources, and Consistent Oversight Needed if Regulatory Analysis Is to Help Control Costs of Regulation* (Washington, D.C.: GPO, November 2, 1982), pp. 50–54. The GAO report is discussed at length in Eads and Fix, *Relief or Reform?*, chap. 6. See also Ludlam, *Undermining Public Protections*.

22. Congressional Quarterly, *Federal Regulatory Directory*, 1981–1982, p. 74.

23. Ibid.

24. Ibid.

25. Thomas D. Hopkins et al., *A Review of the Regulatory Interventions of the Council on Wage and Price Stability, 1974–1980* (Washington, D.C.: CWPS, January 1981), Appendix A; and Murray L. Weidenbaum, "Regulatory Reform Under the Reagan Administration," in George C. Eads and Michael Fix, eds., *The Reagan Regulatory Strategy* (Washington, D.C.: Urban Institute, 1984), p. 24. See also Presidential Task Force on Regulatory Relief, *Reagan Administration Achievements* (Washington, D.C.: GPO, August 11, 1983).

These estimates, like the original cost estimates, should be taken with a grain of salt. The administration took credit for preventing regulations that were unlikely ever to have been issued, and probably exaggerated the costs of proposals that it did prevent.

26. Christopher C. DeMuth, "Constraining Regulatory Costs. Part 1: The White House Review Programs," *Regulation*, January/February 1980, pp. 13–26.

27. "Costle, Muskie Express Views on Cost Cutting in Regulations," *OSH Reporter*, March 1, 1979, p. 1504.

28. In 1976 the Senate Labor Committee challenged the confirmation of Michael Moskow, former head of CWPS, as undersecretary of labor because he had played a leading role in the review program.

29. "Projected Regulatory Analysis Program Expected to Utilize Review Group," *OSH Reporter*, December 29, 1977, pp. 1171–1172.

30. "What Will Happen When the Regulators Regulate Themselves?" *National Journal*, November 4, 1978, p. 1771; "Rogers Asks GAO Investigation of White House Advisors' Role," *OSH Reporter*, June 29, 1978, p. 101; and "Costle, Muskie Express Views," pp. 1504–1505.

31. *Natural Resources Defense Council et al. v. Charles Schultze, Chairman, Council of Economic Advisors et al.*, U.S. District Court for the District of Columbia, Civil Action 79–153, dismissed January 1979.

32. *Sierra Club v. Costle*, Civil Action 79–1565, D.C. Court of Appeals, April 28, 1981.

33. Morton Rosenberg, *Presidential Control of Agency Rulemaking: An Analysis of Constitutional Issues Which May Be Raised by Executive Order 11291*, Committee Print 97-O, Committee on Energy and Commerce, U.S. House of Representatives, 1981, pp. 38, 42. See also Ludlam, *Undermining Public Protections*.

34. Quoted by Ronald Brownstein in "Making the Worker Safe for the Workplace," *Nation*, June 6, 1981, p. 694.

35. "Administration Backs Regulatory Reform Bill," *Congressional Quarterly Weekly Report*, April 11, 1981, p. 627.

36. "Regulatory 'Reform' May Lose to Regulatory 'Revolution' Advocates," *National Journal*, June 14, 1980, pp. 971–972.

37. S. 1080 was first drafted in consultation with the administration and reflected the Reagan position. It included a cost–benefit requirement for independent commissions and executive branch agencies that could be circumvented with OMB approval; it limited court challenges to cost–benefit analyses; and it did not include a legislative veto. But congressional opponents of social regulation succeeded in strengthening the bill in committee, and later drafts included automatic cost–benefit tests, heightened judicial

scrutiny, and the legislative veto. "Regulatory Reform Bill in Logjam," *Washington Post*, November 26, 1982, p. E8; "Administration Backs Regulatory Reform Bill," pp. 627–628.

38. *Sierra Club* v. *Costle*. Also see Michael Sohn and Robert Litan, "Regulatory Oversight Wins in Court," *Regulation*, July/August 1981, pp. 17–24.

39. *American Petroleum Inst.* v. *OSHA*, 581 F.2d 493 (5th Cir. 1978), *aff'd. sub nom. Industrial Union Dep't.* v. *American Petroleum Inst.*, 448 U.S. 607 (1980).

40. *AFL–CIO* v. *Marshall*, 617 F.2d 636 (D.C. Cir. 1979), *aff'd. sub nom. American Textile Mfrs. Inst.* v. *Donovan*, 452 U.S. 490 (1981).

41. "Council on Wage and Price Stability Testimony on Coke Oven Emissions Inflation Impact Statement," *OSH Reporter*, May 13, 1976, pp. 1796–1800; quote is on p. 1799.

42. Ibid.

43. See "Pressure from VP, Study Confirming Benefits Pushed OMB to Issue OSHA Rule," *Inside O.M.B.* 1, no. 7 (March 28, 1982): 1, 5–9.

[7] OSHA

1. Meier, *Regulation*, p. 216.

2. In 1977 and 1979, the following committees considered OSHA activities: the Senate and House Labor Committees; the House Government Operations Committee; the House Select Committee on Small Business; the Senate Commerce, Science and Transportation Committee; the Senate Agriculture, Nutrition and Forestry Committee; the Senate Finance Committee; the Senate Judiciary Committee; the House Science and Technology Committee; the Senate Government Affairs Committee; and the Joint Economic Committee.

3. The average is actually 2 years and 11.2 months. This undoubtedly understates the case; I have used as a measure the time between the first official action by the agency and the date of the standard's publication in the *Federal Register*. But OSHA considers standards before it takes public action. The first official action varied and included the first meeting of a standards advisory committee, the issuance of an emergency temporary standard, advance notice of proposed rulemaking, or notice of proposed rulemaking. In each instance I have used the earliest date. The time taken to develop each of the standards in the table is as follows: asbestos, 6 months; 14 carcinogens, 7 months; vinyl chloride, 6 months; coke-oven emissions, 60 months; benzene, 43 months; DBCP, 6 months; arsenic, 40 months; cotton dust, 42 months; acrylonitrile, 8 months; lead, 37 months; cancer policy, 27 months;

employee access to medical records, 22 months; noise exposure/hearing conservation, 83 months; labeling, 100 months; ethylene oxide, 41 months.

4. For the number of employees covered by inspections, see Office of Technology Assessment, *Preventing Illness*, table A-4; for the number of employees, see U.S. Bureau of the Census, *Statistical Abstract of the United States: 1985*, 105th ed. (Washington, D.C.: Government Printing Office, 1984), table 690, p. 412.

5. U.S. Department of Labor, Occupational Safety and Health Administration, "Federal Compliance Activity Reports," mimeo, various years.

6. Viscusi, *Risk by Choice*, p. 22.

7. SACs included organized labor on an equal footing with industry, and as a result they tended to recommend strict exposure levels. Under the terms of the act, they were also action forcing; the SAC had to make a recommendation, and the agency had to accept or reject it.

8. This standard was developed under Corn's predecessor, John Stender.

9. See "Memorandum to the Under Secretary" from George C. Guenther, June 14, 1972, reprinted in Ashford, *Crisis in the Workplace*, pp. 543–544. As the Nixon White House would have said, the operative section is "While I have discussed with Lee Nunn the great potential of OSHA as a sales point for fund raising and general support by employers, I do not believe the potential of this appeal is fully recognized. Your suggestions as to how to promote the advantages of four more years of properly managed OSHA for use in the campaign would be appreciated."

10. "Corn Tells House Subcommittee That Safety Outweighs Economics," *OSH Reporter*, December 11, 1975, pp. 974–975.

11. "OSHA Will Issue 13 Health Standards by Month's End to Avoid Impact Rules," *OSH Reporter*, September 18, 1975, pp. 731–732; and "OCAW Sues over Inflation Studies; AFL–CIO, IUD Question Delay in Rules," *OSH Reporter*, March 11, 1976, pp. 1331–1332.

12. Eula Bingham, "The New Look at OSHA: Vital Changes," *Labor Law Journal* 29, no. 8 (August 1978): 487–492.

13. "No Obstacle to OSHA Seen from Regulatory Council, Bingham Says," *OSH Reporter*, October 26, 1978, p. 700.

14. "Change in Economic Impact Policy Being Considered, NACOSH Is Told," *OSH Reporter*, March 3, 1977, p. 1253; "Transition Paper on Economic Analysis Not Policy Statement, NACOSH Group Told," *OSH Reporter*, April 14, 1977, p. 1424.

15. "No Obstacle to OSHA," pp. 699–700.

16. The head of the agency's Office of Environmental, Inflationary and Economic Impact announced that he expected "we will find that we have been averting substantially more costs than most people realize." See "$100,000

Contract to Include Study of 'Implicit' Social Costs of Injuries," *OSH Reporter*, August 25, 1977, p. 399. Four peer reviews were ordered in 1978: on labeling, arsenic, lead, and toluene. See "Standard Cost Could Be $14 billion; OSHA to Get Review of Economic Study," *OSH Reporter*, February 23, 1978, p. 1451.

17. Viscusi points out that this jump reflects a decrease in the number of safety inspections. Viscusi, *Risk by Choice*, p. 20.

18. Ibid., pp. 18 and 23.

19. U.S. Department of Labor, OSHA, "Federal Compliance Activity Reports," various years.

20. Office of Technology Assessment, *Preventing Illness*, table A-7, p. 370.

21. Jeffrey Lewis Berger and Steven D. Riskin, "Economic and Technological Feasibility in Regulating Toxic Substances under the Occupational Safety and Health Act," *Ecology Law Quarterly* 7 (1978): 303–308.

22. A "safe" workplace was defined as one in which there were no deaths and a minimum number of workdays were lost because of injury. Although reasonable on its face, the bill allowed employers to redefine the workplace so that separate activities could be treated as separate workplaces. Keep in mind as well that workplace injury records remain a poor guide to health hazards.

23. *Making Prevention Pay, Final Report of the Interagency Task Force on Workplace Safety and Health* (draft), Sec. III, at pp. 15–20, December 14, 1978.

24. "Safety Agency to Forgo 'Cost-Benefit' Analysis," *New York Times*, July 13, 1981, p. A11; and Thorne G. Auchter, "OSHA: A Year Later," *Labor Law Journal* 33, no. 4 (April 1982): 195–201.

25. Auchter, "OSHA: A Year Later," pp. 195–201.

26. Benjamin W. Mintz, *OSHA, History, Law and Policy* (Washington, D.C.: Bureau of National Affairs, 1984), p. 420.

27. Office of Technology Assessment, *Preventing Illness*, table A-5, p. 368; U.S. Department of Labor, OSHA, "Federal Compliance Activity Reports," various years.

28. Michael Wines, "Auchter's Record at OSHA Leaves Labor Outraged, Business Satisfied," *National Journal*, October 1, 1983, p. 2010.

29. Smith, *The OSH Act*, p. 61.

30. Office of Technology Assessment, *Preventing Illness*, table 13-1, p. 259.

31. See "Pressure from VP," pp. 5–9. On industry support, see "Reagan Plan on Labeling Hazards Is Drawing Fire," *New York Times*, October 10, 1983, p. B12; and "State to Enforce Right to Know Law," *New York Times*, December 11, 1983, Sec. 11, p. 1.

32. "U.S. in a Reversal, Maintains Dust Controls at Textile Mills," *New York*

Times, May 20, 1983, p. 16; and "Fight Grows over Cotton Dust Rules," *New York Times*, May 11, 1983, Sec. B, p. 6.

33. Office of Technology Assessment, *Preventing Illness*, p. 235.

34. Viscusi, *Risk by Choice*, pp. 23–24.

35. Ibid., p. 85.

36. Ruth Ruttenberg and Randall Hudgins, *Occupational Safety and Health in the Chemical Industry*, 2d ed. (New York: Council on Economic Priorities, 1981).

37. For a promising effort to splice the two series, see Michele I. Naples and David M. Gordon, "The Industrial Accident Rate: Creating a Consistent Time Series," mimeo, December 1981.

38. Office of Technology Assessment, *Preventing Illness*, pp. 34–36.

39. Smith, "The Impact of OSHA Inspections on Manufacturing Injury Rates," *Journal of Human Resources* 14, no. 2 (Spring 1979): 145–170.

40. David P. McCaffrey, "An Assessment of OSHA's Recent Effects on Injury Rates," *Journal of Human Resources* 18, no. 1 (1983): 131–146.

41. W. N. Cooke and F. H. Gautschi III, "OSHA Plant Safety Programs, and Injury Reduction," *Industrial Relations* 20, no. 3 (1981): 245–257.

42. Mendeloff, *Regulating Safety*, chap. 6.

43. See also W. Kip Viscusi, "The Impact of Occupational Safety and Health Regulation," *Bell Journal of Economics* 10, no. 1 (Spring 1979): 117–140. For summaries of this literature, see James R. Chelius, "The American Experience with Occupational Safety and Health Regulation," *New Zealand Journal of Industrial Relations* 8 (1983): 123–132; and Office of Technology Assessment, *Preventing Illness*, pp. 264–268.

44. Barth and Hunt, *Workers' Compensation*, pp. 15–27.

45. Office of Technology Assessment, *Preventing Illness*, p. 48.

46. John Mendeloff, *An Analysis of OSHA Health Inspection Data*, contract report prepared for Office of Technology Assessment, U.S. Congress, Washington, D.C., April 1983. Cited in Office of Technology Assessment, *Preventing Illness*, p. 268.

47. Office of Technology Assessment, *Preventing Illness*, p. 268.

[8] Regulatory Reform

1. Smith, "Protecting Workers' Health and Safety," pp. 311–338, esp. pp. 321–322; quote is on p. 322.

2. Ibid., pp. 330–334.

3. For an excellent exposition of the economic model underlying this approach, see Dickens, "Occupational Safety and Health Regulation and Economic Theory."

4. Ibid., p. 136.

5. Smith, "Protecting Workers," p. 326.

6. Ibid., pp. 330-334.

7. Ibid., pp. 328-329; and Chelius, *Workplace Safety and Health*, pp. 63-69.

8. Smith, "Protecting Workers," p. 327.

9. Chelius, *Workplace Safety and Health*.

10. Dickens, "Occupational Safety," pp. 134-135, 137-138, 141, 147-148. On the marginal worker's characteristics, see Freeman and Medoff, *What Do Unions Do?*, pp. 9-11.

11. Smith, "Protecting Workers," p. 321.

12. Michal S. Baram, *Alternatives to Regulation* (Lexington, Mass.: Lexington Books, 1982), p. 81.

13. For the best statement of this approach, see Stephen Breyer, *Regulation and Its Reform* (Cambridge: Harvard University Press, 1982). See also Robert E. Litan and William D. Nordhaus, *Reforming Federal Regulation* (New Haven: Yale University Press, 1983).

14. Lester B. Lave, *The Strategy of Social Regulation* (Washington, D.C.: Brookings Institution, 1981), pp. 23-25.

15. Zeckhauser and Nichols, *OSHA*, p. 226.

16. Lave, *Strategy*, pp. 19-21.

17. DeMuth, "Constraining Regulatory Costs. Part 1," p. 26; idem, "Part 2: The Regulatory Budget," *Regulation*, March/April 1980; and Litan and Nordhaus, *Reforming Federal Regulation*, chap. 6.

18. Zeckhauser and Nichols, *OSHA*, pp. 188-191. Though the authors prefer performance standards to detailed physical standards, they have doubts about the degree to which performance standards will produce an optimal level of efficiency compared to a reformed compensation system.

19. Schultze, *The Public Use of the Private Interest*.

20. The general argument for incentive mechanisms is discussed in Steven Kelman, *What Price Incentives* (Boston: Auburn House, 1981); on performance standards, see Viscusi, *Risk by Choice*, pp. 128-132; on OSHA, see *Making Prevention Pay*, chap. 4, pp. 14-17.

21. E. J. Mishan, *Cost-Benefit Analysis* (New York: Praeger, 1976), p. 414.

22. Viscusi, *Risk by Choice*, p. 82.

23. Ibid., p. 80.

24. Office of Technology Assessment, *Preventing Illness*, pp. 78-85.

25. The classic piece remains Philippe C. Schmitter, "Still the Century of

Corporatism?" *Review of Politics* 36 (1974): 85–131. For a more critical view, see Leo Panitch, "The Development of Corporatism in Liberal Democracies," *Comparative Political Studies* 10, no. 1 (April 1977): 61–90.

26. See Gevers, "Worker Participation," pp. 413–414. See also Les Boden and David Wegman, "Increasing OSHA's Clout: Sixty Million New Inspectors," *Working Papers for a New Society* 6 (May/June 1978): 43–49.

27. Leopold and Beaumont, "Joint Health and Safety Committees in the United Kingdom," pp. 270–271; Glendon and Booth, "Worker Participation in Occupational Health and Safety in Britain," p. 400; and Gevers, "Worker Participation," p. 422.

28. Gevers, "Worker Participation," pp. 415–417.

29. Leopold and Beaumont, "Joint Health and Safety Committees in the United Kingdom," pp. 270–271, 278–280; Glendon and Booth, "Worker Participation in Occupational Health and Safety in Britain," pp. 400–406; and Gevers, "Worker Participation," pp. 412, 416–419, 422, 424.

30. Hauss and Rosenbrock, "Occupational Health and Safety in Germany," p. 284; and Gevers, "Worker Participation," pp. 420–421.

31. Hauss and Rosenbrock, "Occupational Health and Safety in Germany," pp. 279–284; Gevers, "Worker Participation," p. 421; and Matt Witt and Steve Early, "The Worker as Safety Inspector," *Working Papers*, September/October 1980, pp. 21–29.

32. Gevers, "Worker Participation," pp. 417–425. For an interesting brief for a collective-bargaining approach in the United States, see Lawrence S. Bacow, *Bargaining for Job Safety and Health* (Cambridge: MIT Press, 1980).

33. Deutsch, "Work Environment Reform," pp. 180–194, quote on p. 187; Eiger, "Economic Democracy and the Democratizing of Research," pp. 123–141; Kelman, *Regulating America, Regulating Sweden*, passim; and Gardell, "Scandinavian Research on Stress in Working Life," pp. 31–41.

34. Kelman, *Regulating America, Regulating Sweden*.

35. Deutsch, "Work Environment Reform," p. 187; Parmeggiani, "State of the Art," p. 276; Frank Goldsmith, "Sweden's New Occupational Health Law: The Impact on Occupational Physicians," *Journal of Occupational Medicine* 21, no. 11 (November 1979): 761.

36. On the rest, the agencies had set the same levels. This analysis is based on my comparison of TWAs from the "Alphabetical List of Substances" in the International Labour Office's *Occupational Exposure Limits for Airborne Toxic Substances*, 2d ed. I have restricted this comparison to substances that both countries regulate. Differences in the economies of these two nations are likely to yield differences in the use of toxic substances and, consequently, different health hazard priorities. See International Labour Office, *Occupational Exposure Limits for Airborne Toxic Substances*, 2d ed., Occupational

Safety and Health Series no. 37 (Geneva: ILO, 1980). Steven Kelman does not consider these data, and his study of rulemaking in Sweden reaches different conclusions. He stresses the "surprisingly similar" regulations of OSHA and ASV and attributes these similarities to the strategic role that protectionist professionals play in both agencies. But Kelman's evidence actually supports my conclusion. He does note that "the rule-making decisions on occupational safety and health in the United States and Sweden during the period considered were very similar but not identical, and differences that did exist were in the direction of higher protection in Sweden" (p. 110). Small differences in occupational safety and health standards are often very important. Kelman identifies four specific instances of rulemaking in which ASV chose a more protective standard than OSHA: the retrofitting of tractors with rollover protective devices; the installation of load-indicating devices on cranes; the banning of asbestos cement; the establishment of an 85-dBA noise standard. Kelman also notes that ASV established lower TLVs for a number of toxic chemicals (p. 81). Kelman, *Regulating America, Regulating Sweden*.

37. Office of Technology Assessment, *Preventing Illness*, p. 261; Kelman, *Regulating America, Regulating Sweden*, passim.

38. Glendon and Booth, "Worker Participation," pp. 409–410.

39. The works council's authority is statutorily limited because members are a minority on the plant health and safety committee. It generally confines its deliberations to organizational matters such as the selection and functioning of health experts. See Gevers, "Worker Participation," pp. 420–421; Hauss and Rosenbrock, "Occupational Health and Safety in Germany," p. 284.

40. The literature on the problems and prospects is enormous and growing. See Walter Korpi, *The Democratic Class Struggle* (London: Routledge & Kegan Paul, 1983); John Stephens, *The Transition from Capitalism to Socialism* (London: Macmillan, 1979); Andrew Martin, "The Dynamics of Change in a Keynesian Political Economy: The Swedish Case and Its Implications," in Colin Crouch, ed., *State and Economy in Contemporary Capitalism* (New York: St. Martin's Press, 1979), pp. 88–121; Jonas Pontusson, "Behind and Beyond Social Democracy in Sweden," *New Left Review* 143 (January–February 1984): 69–96; Mary Nolan and Charles F. Sabel, "Class Conflict and the Social Democratic Reform Cycle in Germany," in Zeitlin, ed., *Political Power and Social Theory*, vol. 3 (1982), pp. 145–173; Mark Kesselman, "Prospects for Democratic Socialism: Class Struggle and Compromise in Sweden and France," *Politics and Society* 11, no. 4 (1982): 397–438; Gosta Esping-Andersen, *Politics Against Markets: The Social Democratic Road to Power* (Princeton: Princeton University Press, 1985); and Adam Przeworski, "Social Democracy as an Historical Phenomenon," *New Left Review* 122 (1980): 27–58.

[9] Conclusion

1. The classic statement of this phenomenon remains Francis Fox Piven and Richard A. Cloward, *Regulating the Poor: The Functions of Public Welfare* (New York: Vintage Books, 1971). See also Nolan and Sabel, "The Social Democratic Reform Cycle."

2. On the enduring liberalism of the American electorate, see William Watts and Lloyd A. Free, *State of the Nation III* (Lexington, Mass.: Lexington Books, 1978), pp. 87–95; Seymour Martin Lipset, "The Economy, Elections, and Public Opinion," *Tocqueville Review*, Fall–Winter 1983, pp. 431–469.

3. See, for example, Schwarz, *America's Hidden Success*; Thomas Bryne Edsall, *The New Politics of Inequality* (New York: Norton, 1984).

Interview Sources

Dr. Martin Anderson, Domestic Policy Adviser, Nixon administration
Stanley Aronowitz, Oil, Chemical, and Atomic Workers
Frank Barnako, Republic Steel and Occupational Safety and Health
Review Commission
Monroe Berkowitz, Chair of President's Task Force on Workmen's
Disability Income
Richard Burress, Deputy Counsel to the President and Counsel to
Arthur Burns, Nixon administration
Joseph Califano, Domestic Policy Adviser, Johnson administration
George Cohen, Bredhoff and Kaiser
Wilbur Cohen, Secretary of Health, Education, and Welfare, Johnson
administration
Representative Dominick Daniels (D-N.J.)
Barbara Geller, aide to Dr. Eula Bingham
Frank Greer, OSHA, Carter administration
Louis Hastings, Automobile Manufacturers Association
James Hodgson, Secretary of Labor, Nixon administration
P. Sam Hughes, Assistant Director for Legislative Clearance and
Deputy Director, Bureau of the Budget, Johnson administration
Senator Jacob Javits (R-N.Y.)
Paul Jensen, Department of Labor, Carter administration
Simon Lazarus, Associate Director, Domestic Policy Staff, Carter
administration
Dr. Phillip Lee, Public Health Service, Johnson administration
Lawrence Levinson, labor liaison, Johnson White House Staff
Anthony Mazzochi, Oil, Chemical, and Atomic Workers
Eugene Mittelman, Minority Counsel, Senate Labor Committee, and
aide to Senator Javits

Len Mote, Automobile Manufacturers Association

Robert Nagle, Senate Labor Committee staff and aide to Senator Harrison Williams

Alfred Neal, President of the Committee for Economic Development

Dr. Norton Nelson, Chair, Task Force on Occupational Health, Public Health Service, Johnson administration

Anthony Obadal, U.S. Chamber of Commerce

Esther Peterson, Assistant Secretary of Labor, Johnson administration

Ruth Ruttenberg, OSHA economist, Carter administration

Richard Schubert, aide to Secretary of Labor James Hodgson, Nixon administration

Charles Schultze, Director of the Bureau of the Budget under Johnson and Chair of the Council of Economic Advisers under Carter

Margaret Seminario, AFL–CIO

John J. Sheehan, Washington Legislative Director, United Steel Workers of America

Lawrence Silberman, Solicitor of Labor, Nixon administration

Dr. James Sterner, Public Health Service Council on Environmental Health

David Swankin, Director, Labor Standards Bureau, Johnson administration

George Taylor, Health and Safety Director, AFL–CIO

Leo Teplow, American Iron and Steel Institute

Frank Wallich, United Auto Workers

Dr. Mary Ellen Weber, Director, OSHA Office of Regulatory Analysis, Carter and Reagan administrations

Senator Harrison Williams (D-N.J.)

W. Willard Wirtz, Secretary of Labor, Johnson administration

Steven Wodka, Oil, Chemical, and Atomic Workers

The following people responded in writing:

Dr. Arthur Burns, Director of Nixon Transition Task Force

John Gardner, Secretary of Health, Education, and Welfare, Johnson administration

Reginald Jones, General Electric

Alexander Trowbridge, Secretary of Commerce, Johnson administration

Acronyms

ACGIH	American Conference of Governmental Industrial Hygienists
AEC	Atomic Energy Commission
AFL	American Federation of Labor
AFL-CIO	American Federation of Labor-Congress of Industrial Organizations
AIHA	American Industrial Hygiene Association
AIHC	American Industrial Hygiene Council
AISI	American Iron and Steel Institute
AMA	American Medical Association
ANSI	American National Standards Institute
ASSE	American Society of Safety Engineers
ASV	Arbetarskyddsverket (Sweden)
BC	Business Council
BLS	Bureau of Labor Statistics
BOB	Bureau of the Budget
BOSH	Bureau of Occupational Safety and Health
CEA	Council of Economic Advisers
CED	Committee for Economic Development
CEQ	Council on Environmental Quality
CIO	Congress of Industrial Organizations
COH	Council on Occupational Health
COSH	Committees, Councils, Coalitions or Projects on Occupational Safety and Health
CPSC	Consumer Product Safety Commission
CWPS	Council on Wage and Price Stability

DOL	Department of Labor
EFFE	Environmentalists for Full Employment
EO	Executive Order
EOP	Executive Office of the President
EPA	Environmental Protection Agency
EPB	Economic Policy Board
EPG	Economic Policy Group
FRC	Federal Regulatory Council
FTC	Federal Trade Commission
GAO	General Accounting Office
HEW	U.S. Department of Health, Education, and Welfare
HRG	Health Research Group
IAGLO	International Association of Governmental Labor Organizations
IAIABC	International Association of Industrial Accident Boards and Commissions
IAM	International Association of Machinists
IHF	Industrial Hygiene Foundation
IIS	Inflationary Impact Statement
IMA	Industrial Medical Association
IUD	Industrial Union Department, AFL–CIO
LO	Landsorganisationen (Sweden)
LSB	Labor Standards Bureau
NAM	National Association of Manufacturers
NCF	National Civic Federation
NCFE	National Committee for Full Employment
NCLC	National Chamber Litigation Center
NIOSH	National Institute for Occupational Safety and Health
NLCPI	National Legal Center for the Public Interest
NLRB	National Labor Relations Board
NRDC	Natural Resource Defense Council
NSC	National Safety Council
OCAW	Oil, Chemical, and Atomic Workers
OIRA	Office of Information and Regulatory Affairs
OMB	Office of Management and Budget
ORC	Organizational Research Counselors
OSHRC	Occupational Safety and Health Review Commission
OTA	Office of Technology Assessment

PEL	permissible exposure level
PHS	Public Health Service
PPD	personal protective device
QES	Quality of Employment Survey
RARG	Regulatory Analysis Review Group
RC	Regulatory Council
RIA	Regulatory Impact Analysis
SAC	Standard Advisory Committee
SAF	Svenska Arbetsgivareföreningen (Sweden)
SPI	Society of the Plastics Industry
TCO	Tjänstemannens Centralorganisationen (Sweden)
TLV	threshold limit value
TUC	Trades Union Congress (Great Britain)
UAW	United Auto Workers
UEC	Urban Environment Conference
UMW	United Mine Workers
URW	United Rubber Workers
USWA	United Steel Workers of America

Index